Science AND Societal Issues

A GUIDE FOR SCIENCE TEACHERS

CHARLES R. BARMAN

JOHN J. RUSCH

TIMOTHY M. COONEY

The Iowa State University Press
AMES, IOWA

Charles R. Barman is an associate professor at Buena Vista College.

John J. Rusch is an associate professor at the Center for the Advancement of Science, University of Wisconsin-Superior.

Timothy M. Cooney is an assistant professor at the Malcolm Price Laboratory School, University of Northern Iowa.

Printed by
The Iowa State University Press
Ames, Iowa 50010

First edition, 1981

International Standard Book Number: 0-8138-0485-X

Library of Congress Catalog Card Number: 81-80205

CONTENTS

PREFACE

The world your students are experiencing is quite different from the world you knew as a child and adolescent. As a child you read about space travel only in science fiction. But this mode of transportation was a reality long before your students were born.

Extensive changes have taken place in our society within the last few decades. As changes occur, your students will be faced with possible modifications in their attitudes and values. For example, it seems likely that your students may be forced to decide how and when they will die. There also appears to be good evidence that during their lifetimes, humans may make great strides toward controlling their own evolution.

The changes in science and technology that our students will witness will demand that they have a greater awareness and input into science and public policy issues. These considerations should force all science educators into looking at existing science curricula to determine whether they meet the future needs of their students. It will be imperative that science curricula provide opportunities for students to interact with their peers and adults regarding science topics that will have a personal impact on them.

The chapters and activities of this book have been designed to provide you with a rationale and a model for integrating values issues in science teaching and to aid you in coping with controversial topics in your classroom.

Science and Societal Issues

Firemen End Strike
FBI Agent Indicted
TURN US ON. WE'LL TURN YOU ON.
PRES. KENNEDY ASSASSINATED
National Smoker Study Hails Merit.
Good taste.
HOW TO GET RICH THE LAZY MAN'S WAY
Wisconsin, you certainly know how to enjoy a holiday.
Child abuse hurts everybody.
Reward Is Offered for Kidnapped Son
Terrorist Seized In Germany
The Homosexual Teacher
Ugandan Invasion Advances
School Talks Are Ordered Resumed

1

VALUES AND YOUR STUDENTS

Your students' lives are filled with a variety of experiences, such as the ones illustrated on the preceding page. From these experiences they sort out what it is they like and dislike. They develop either positive or negative attitudes toward certain things. Eventually, these attitudes may develop into general guides to their behavior. These guides tend to direct their lives and are referred to as values.[1]

Today, students have a more difficult time selecting what it is about their lives that is truly of value. As members of the "Affluent Society," they are offered an inexhaustible selection of consumer goods and services. Trying to make a decision about these goods and services is an experience that could place a person's mind into "future shock." For example, a person walking into a supermarket has not only to decide what products to buy, but also which brand name to select. In many cases, this decision is made more difficult by comparisons between the regular size and the economy size or between volume or mass and price.

A person confronted with making a decision regarding the wide variety of services in our society could also find that he may be in a similar state of confusion. Decisions concerning whose services to hire are extremely difficult. This decision conflict becomes quite apparent when you look through the "yellow pages" of a telephone book to select the services of a specific business. An initial response may be to "flip a coin" and rely on chance to meet your needs. However, the consequences of this decision-making technique could be economically unwise.

It is understandable that decisions that may have been simple choices fifty years ago may no longer be viewed this way. It is also understandable that these now complicated decisions add to our anxiety and frustration.

Attempting to decide what is "good" and "bad" behavior is equally difficult. Almost every day our students view and hear via mass media that our world is full of social disorder and unrest. They are told not to break the laws of our society, yet they hear that some governmental officials and corporations have not followed these rules. Many movies and books display various aspects of deviant behavior in our society, such as murder, rape, and stealing. Is it any wonder that young people today find themselves in real conflict when attempting to resolve what they should and should not value?

Two other aspects contributing to the development of our students' values are the family and the school. (We realize there are many other experiences that also influence a student's value development, such as peer groups, religious education, etc. To list all of these would be virtually impossible.) The family was and still is the most important influence on a person's valuing system.[2] However, the structure of the family has changed drastically in the last few decades. This is exemplified by the high rates of divorce, separation, and one parent families. Family members tend to share less time together and spend more time doing "their own thing." Therefore, the family influence on each individual's values has decreased in most cases.

Schools also have an impact on the value development of our students. These institutions have established rules that tell students what the administration considers to be proper behavior. In addition, teachers have their own values and attitudes. These are reflected by their opinions and behavior which definitely influence their students' thoughts and actions. In other words, schools are not and cannot be value free.

An Example of How Schools Can Influence a Student's Values

Situation I

Today your chemistry class will be conducting a laboratory activity. The investigation you have planned requires the use of expensive glassware. During the second period, you observe John Doe tossing and returning a yo-yo while working on the investigation. You direct John to put the yo-yo on your desk. John is reluctant, but says he will put it away. Again, you tell John to put the yo-yo on your desk. John indicates that he is afraid you will not return the yo-yo after class. You inform John of two choices; either he puts the yo-yo on your desk or he must leave class. Knowing that if he leaves class he will receive no credit for the lab activity, John gives in and turns over the yo-yo.

When second period ends, John asks you to return his yo-yo. How would you respond to John?

Situation II

AS A TEACHER, WHAT WOULD YOU DO?

Situation III

A teacher in a ninth grade science class asks the students to prepare, as part of their normal academic requirements, a quarterly book report. At the end of the third quarter, while grading the book reports, the teacher finds two reports that are identical. However, there are two different names on them.

The teacher places comments on both of these reports, noting their similarity and requesting that the two students confer with her. No grades are given on these reports.

One of the students immediately comes to the teacher and indicates that she did not copy the book report; and that indeed as evidence of this, she can produce an eighth grade report from which she copied this one.

The teacher responds to the student that it is difficult to grant a grade on a book report submitted in a ninth grade class that was already used for credit in an eighth grade class. The teacher then tells the student that she will be given a lower grade on this report, due to these circumstances.

During the third quarter parent-teacher conferences, the mother and the student referred to above meet with the teacher. They discuss the situation concerning the lowered grade.

The mother tells the teacher that she broke the trust with her daughter. After all, her daughter went to her honestly and told her that she had used the book report in eighth grade.

The teacher indicates to the mother that it would be difficult, in her mind, to grant credit for work previously submitted in another class. The mother is adamant and, while her daughter is present, tells the teacher that she will accept the responsibility of her daughter's grade and demands the grade be changed.

What should the teacher do? Why?

Situation IV

HOW WOULD YOU RESPOND TO THIS SITUATION?

These four situations have been designed to show you how a teacher can inadvertently teach an attitude or value. Whatever decision or action you selected indicates your values or attitudes for that specific situation. Your responses could tend to influence the behavior of your students in the future, even though you did not make an attempt to teach any particular value or attitude. This is why we contend that schools cannot be value free.

Therefore, if educators do influence the value choices of their students intentionally and unintentionally through a teacher's and administrator's rules, words, and actions, we believe that it is essential that time be allotted for students to examine their own beliefs. Value development is a very personal process that is filled with much conflict and frustration; it requires time for a person to think through values decisions alone or within the context of group interaction.

Schools have a tendency to concentrate on teaching facts and concepts and, in many cases, have attempted to stay clear of the affective domain. In some cases where schools have paid some attention to certain values and attitudes, it has been by a process of inculcation. This approach will do nothing to promote value decision making. Values decisions must be made freely by considering various alternatives before a final selection is formulated. Therefore, we support an educational methodology that would allow students to draw their own conclusions on what it is they value. This type of instruction would allow the students to freely interact with their peers and adults in a discussion of issues that have particular import to them. It would also provide the opportunity to gain insight into the fact that many values issues can be viewed differently and can have more than one answer.

Experiences of this type add another dimension to a student's value development. Students are given the opportunity to freely explore certain values issues in the classroom, as well as outside of school, thus providing more relevance to their education.

The activities in the last portion of this book require students to make values decisions about science and social issues. Before using these lessons, we suggest you read the remaining chapters of this book. You will then acquaint yourself with additional information on values and value development and how these relate to science education.

REFERENCES

[1] Raths, Louis E., et al., Values and Teaching, Charles E. Merrill Publishing Co., Columbus, Ohio, 1966.

[2] Ibid.

2

WHAT ARE VALUES ALL ABOUT?

In Chapter 1 we listed some of the experiences that students may encounter which could contribute to the development of their values. Now, let's take a closer look at how values appear to be formed.

Krathwol, et al., have devised a taxonomy of value development which they refer to as the affective domain.[1] Their classification system includes five main categories. However, for our purposes we will condense this system into three general groups:

1. Awareness—A person is aware of and responds to a stimulus in his environment (i.e., a person is willing to attend school).

2. Acceptance of values—The person decides that school is important and makes a commitment to study.

3. Preference for values—The person has formulated a general guide that school is beneficial to him. He has integrated these attitudes into a philosophy of life.

Values are formed by our daily experiences; therefore, it seems reasonable to assume that people from different parts of the United States or the world would have different values. It also seems reasonable that if a person has a modification in his daily experiences a change could occur in his values.

Because values mold our behavior, they are developed through extremely complex circumstances. We are placed into situations that cause us to re-examine what it is we feel is right or worth believing. Therefore, a value is not static. It is something that is constantly being shaped by our daily experiences.

Developing Values

According to Louis E. Raths, et al., a person formulates a value through a valuing process.[2] It is their contention that a value is developed through the following steps:

1. Choosing—The person must be able to select what he perceives to be good or bad freely, from a series of alternatives, and after thoughtful consideration of the consequences of each alternative.

2. Prizing—After the person makes his selection, he must be satisfied with it. He may demonstrate his satisfaction by publicly affirming his choice.

3. Acting—The person must act upon the choice with some repetition.

For example, let's say Sue Smith feels very strongly that everyone should strive to conserve energy. She realizes that to do this it will take some personal discomfort. However, she feels the other alternatives are less satisfactory. Sue explains her position to her friends. She also uses the public transportation system, keeps her thermostat set at 65 degrees F., and walks whenever possible. One could say that Sue Smith has developed the value that the energy problem is everyone's concern.

Values Are Personal

As stated earlier in this chapter, values are a result of personal experiences and freedom of choice. However, these experiences are influenced by the environmental conditions in which a person lives. Someone who is raised in Nebraska will most likely develop a different value about "getting away from it all" than a person living near the Rocky Mountains. The person from Nebraska may feel "free" on the open prairie but extremely enclosed in a heavily wooded mountain area. On the other hand, a mountain person accustomed to living near a vast number of trees may feel free in a wooded area, provided no other people are nearby.

In the example cited above, it would be impossible to judge which person's values are "right." They are derived from different frames of reference and would be considered very personal choices. If we were to draw a parallel between the last example and society, we might also gain insight into the complexity of value development in a country like the United States.

The United States has great geographical differences as well as ethnic and cultural variances. Add to these differences a tremendous mobility and it becomes quite evident that people in our community will hold some different values. Because this diversity of values exists at home and across the United States, it seems reasonable to assume that our schools should not be judgmental about what values their students possess. Instead, the schools should be sensitive to the fact that valuing is a personal lifelong process that will continue to be altered from childhood through adulthood. Please do not infer from this that we believe educators should ignore values in the classroom. This is impossible. A teacher's attitudes and values are constantly showing through his or her words and actions. However, these are the teacher's personal values and it may be possible that they are in conflict with those of his or her students. Therefore, a teacher should be very careful not to impose certain beliefs on his or her students. Instead, students should be provided with value decision-making experiences and time to clarify their own value systems.

Values, Morals, Morality, and Ethics

It may be useful, at this point, to differentiate between the terms values, morals, morality, and ethics. As stated before, a value is a collective term for what a person decides is good or bad. For our purposes, we will assume that a person's values can be divided into two main categories: (l) nonmoral - a person's evaluation of what is good or bad with respect to objects such as a car, a painting, or a type of food. (2) moral - a person's assessment of what would be considered good or bad actions and/or motives with respect to others.[3] Stated another way, moral values are those that carry an obligation or responsibility to one or more people. When a person or a group of people decide what their moral values are, these are then referred to as the individual's or group's morality. In other words, morality is a collective term for the moral values held by an individual or by a group of individuals.

For example, murder is a moral value that is considered wrong by most individuals of our society. This would appear to be a rather easily defined moral value. However, anyone who reads the newspaper or listens to the radio or TV is aware that there is much controversy regarding this moral value. In fact, our society has categorized this moral value into various degrees, first degree murder being more wrong than second or third degree. This analysis or interpretation of a moral value is referred to as the field of ethics. Ethics, then, deals with the thinking about the degree of rightness or wrongness of morality, moral problems, and moral decisions.[4]

Earlier, we stated that people are constantly modifying their value systems based on specific experiences they encounter. The portion of this continual modification that deals with moral values could be referred to as "moral development." Lawrence Kohlberg has studied this progression for a number of years. The last section of this chapter will deal with Kohlberg's Moral Development Theory that was derived from this research.

Kohlberg's Model of Moral Development

Kohlberg believes that a person's moral development is a sequential process. His model consists of three levels, each containing two stages. It is Kohlberg's contention that a person cannot reach one stage before experiencing a preceding one. However, this person could operate at one stage and revert back to a lower one in certain situations. In addition, Kohlberg's research indicates that people in all cultures progress through the same sequence.[5] (Although Kohlberg does not attach any ages to his levels, we have attempted to do so to make these stages more useful for teachers. These have been derived from data collected by Kohlberg, in which he has graphed the ages of the subjects he tested versus their responses to various moral reasoning situations.[6] These responses were compared with the characteristics of the following six stages to determine the level of moral development of each person. These ages are in no way an accurate assessment of all people and should only be used as a guide.)

The six stages of Kohlberg's model are presented in Figure 2-1 on pages 11-15.

FIGURE 2-1

Preconventional Level	Stage I (Avoid Punishment)	Examples of Stages
(Ages 7-10; Moral decisions are based on self-centered needs.)	A person's actions are based on a blind obedience to avoid punishment or to seek a reward. The "ultimate wrong" is getting into trouble.	1. A child puts his toys away so he may go outside to play with his friends. 2. A child puts her toys away to avoid being punished by her parents.
	Stage II (Self-Benefit) A stage two person is aware of the needs and motives of others and feels "one good or bad turn deserves another."	1. A person buys a treat for a friend with the intention of receiving a favor in return. 2. The other day Jim was reprimanded by his biology teacher for fooling around in class. While walking home from school Jim spots this teacher's car parked on the street. Jim decides to deflate two of the car's tires to get back at his teacher.

FIGURE 2-1 (continued)

Conventional Level	Stage III (Concern for Acceptance)	Examples of Stages
(Ages 10-13. Characterized by conformity and the maintenance of "law and order.")	These individuals are extremely sensitive to peer expectations and approval. This is the first time that behavior is judged on the basis of intention.	1. Susie Jones wants to go to the movies, but her parents say no. She in turn argues that "all her friends will be there." 2. The Smiths felt a need to buy a new car because their friend had recently purchased one.
	Stage IV (Authority Orientation) A stage four person is motivated to act according to what is "right or best" for society. His actions are highly influenced by rules, authority figures, and preserving social order for its own sake.	1. A child is told by his parents to be home at a specific time. He responds to this order without question. The parents may or may not have a good reason for demanding he follow their orders, but the child's action is based on a respect for their authority. 2. A person is asked why he always drives his car at the speed limit. He responds by saying, "That's the law."

Postconventional Level (Ages 13-adult; Decisions are based on values shared by others rather than self-centered or blind interests.)	Stage V (The Greatest Good for the Greatest Number) These individuals feel that the rights of others should be respected and that morality is a matter of personal choice. These people feel that laws should be changed when they no longer aid society. Stage V people would prefer to change a law rather than break it. However, if the modification is not possible, they may choose to break it and willingly suffer the consequences.	1. Throughout the history of our country some young men have resisted induction into the armed forces. They did so on the basis that the specific war they were being forced into was immoral. They were fully aware that their resistance could cause them to be imprisoned. However, they felt strongly enough about their convictions to cause them to disobey their induction order and suffer the consequences. 2. A student confides in Mr. Jones about his recent involvement in the theft of a math test. A day later in the faculty lounge, Mr. Jones hears that three students are going to receive a failing grade in math due to their confessed guilt in the theft of a math test. However, the student who confided in him was not one of the three. Mr. Jones reasons that revealing this student's involvement would do

Examples of Stages

nothing to develop the student's moral character and would completely destroy his rapport with this student. Therefore, Mr. Jones decides to discuss this issue with the student, but allow the student to make his own decision regarding this matter. (In this situation, Mr. Jones believes that a person cannot be coerced into morality, but rather it is a matter of a personal choice.)

Stage VI (Conscience is the Guide)	Examples of Stages
Stage six people feel that individuality is primary to moral decision-making. "Right" is defined by a person's conscience based on self-selected ethical principles. These are universal principles, such as justice, equality of human rights, and the respect for the dignity of human beings as individuals.	1. Dr. Martin Luther King felt compelled to break some of society's laws because he felt they were unjust in dealing with human rights. His conscience directed him to break these laws because he felt his actions were consistent with certain universal ethical principles. 2. The fathers of our country signed the Declaration of Independence to sever the American colonies' ties with England. This document was founded on the basis that all men were created equal. Even though this action was viewed by England as illegal and punishable by death, these men refused to follow England's laws and instead were guided by their conscience. Their behavior could be specific self-selected ethical principles.

Expanding Kohlberg's Model

Kohlberg has also theorized that a person's moral development may extend into a seventh stage.[7] In this stage a person would transcend from a human to a cosmic point of view. It would appear that a stage seven person would experience life similar to what Abraham Maslow calls self-actualization and proponents of Transcendental Meditation refer to as cosmic consciousness.[8]

To better describe the way a stage seven person would view his existence, let's use the following quote that explains cosmic consciousness.

> " . . . the person in cosmic consciousness can make decisions according to ultimate values rather than shortsighted needs. . . . Sustained unbounded awareness enables such an individual to realize a natural sense of morality, justice, beauty and truth, the values projected by Maslow as the deepest level of man's nature."[9]

We feel that it is appropriate to point out that Kohlberg's Theory implies that there are "universal principles." There are some people who disagree with this premise.[10] However, we feel that Kohlberg's Theory has utility in terms of describing human moral behavior and as a model to explain how moral behavior is acquired.

Kohlberg's Theory and American Society

Research studies indicate that 70 percent of all Americans make most of their moral decisions based on the Conventional Level.[11] Therefore, it would follow that only a small minority operate out of a Postconventional frame of reference. It could also be stated, from this data, that the majority of moral decisions in the United States are based on conformity and the respect for authority and law and order. This should give educators a clue as to the fundamental framework that directs most of the moral decisions of their students. In addition, one could infer that if 70 percent of all Americans base their moral decisions on the Conventional Level, the moral development of most students never progresses past stage four.

Kohlberg believes that through a specific values education methodology, teachers can assist students in a continual moral growth. More will be said about this in Chapter 3.

REFERENCES

[1]Krathwol, David, et al., Taxonomy of Educational Objectives, The Classification of Educational Goals, Handbook II: Affective Domain, David McKay, New York, 1964.

[2]Raths, Louis E., et al., Values and Teaching, Charles E. Merrill Publishing Co., Columbus, Ohio, 1966.

[3]Frankena, William K., Ethics (2nd ed.), Prentice-Hall, Inc., Englewood Cliffs, New Jersey, 1973.

[4]Ibid.

[5]Mattox, Beverly A., Getting It Together, Pennant Press, San Diego, California, 1975.

[6]Kohlberg, L., Developmental Psychology Today, CRM Books, Del Mar, 1974.

[7]Mattox, Beverly A., Getting It Together, Pennant Press, San Diego, California, 1975, p. 46.

[8]Bloomfield, H. H., et al., TM Discovering Inner Energy and Overcoming Stress, Dell Publishing Co., Inc., New York, N.Y., 1975.

[9]Ibid. p. 202.

[10]Purpel, David and Ryan, Kevin (eds.) Moral Education (Parts III & IV), McCutchan, Berkeley, CA, 1976.

[11]Mattox, Beverly A., Getting It Together, Pennant Press, San Diego, California, 1975, p. 22.

3

VALUES EDUCATION AND THE SCHOOLS

According to a Gallup Poll, values education is now favored by four out of five Americans.[1] However, the term *values education* has different meanings for different people. Some individuals view values education as the instilling of specific values into students, while others see it as a process in which students examine their beliefs and formulate their own judgments on specific moral and nonmoral decisions.

Douglas Superka, et al., have compiled an overview of the main values education approaches used in American schools.* This overview is as follows:

Five Values Education Approaches

Approaches	Purposes	Methods
Inculcation	-to instill or internalize certain values in students -to change the values of students so they more nearly reflect certain desired values	modeling, positive and negative reinforcement, mocking, nagging, manipulating alternatives, providing incomplete or biased data, games and simulations, role playing, discovery learning
Moral Development	-to help students develop more complex moral reasoning patterns based on a higher set of values -to urge students to discuss the reasons for their value choices and positions, not merely to share with others, but to foster change in the stages of reasoning of students	moral dilemma episodes, with the small group discussion being relatively structured and argumentative

*Reprinted with permission of Social Science Education Consortium, Inc., Boulder, Colorado 80302.

Approaches	Purposes	Methods
Clarification	-to help students become aware of and identify their own values and those of others -to help students communicate openly and honestly with others about their values -to help students use both rational thinking and emotional awareness to examine their personal feelings, values, and behavior patterns	role-playing games, simulations, contrived or real value-laden situations, in-depth self-analysis exercises, sensitivity activities, out-of-class activities, small group discussion
Action Learning	-those purposes listed for analysis and clarification -to provide students with opportunities for personal and social action based on their values -to encourage students to view themselves as personal social interactive beings, not fully autonomous, but members of a community or social system	the methods listed for analysis and clarification as well as action projects within the school and community and skill practice in group organizing and interpersonal relations
Analysis	-to help students use logical thinking and scientific investigation to decide value issues and questions -to help students use rational, analytical processes in interrelating and conceptualizing their values	structured rational discussion that demands the application of reasons as well as evidence, testing principles, analyzing analogous cases, debate, and research

How Do You Feel About Values Education?

To assess your attitudes about values education, we have included the following self-evaluation. This exercise will give you some insight into which values education approach(es) you feel is/are appropriate for your classroom.

Indicate whether you agree with, disagree with, or are not sure about the following statements by checking the appropriate space on the right.

I feel a values education approach should:

		Agree	Disagree	Not Sure
1.	assist students in becoming aware of their own values	______	______	______
2.	change the values of some students so they more clearly reflect society's values	______	______	______
3.	provide students with experiences to act on their values	______	______	______
4.	assist students in using logical thinking in decision-making	______	______	______
5.	urge students to interact with one another to discuss reasons for their value choices	______	______	______
6.	instill positive values in students	______	______	______
7.	assist students in developing more complex moral reasoning patterns	______	______	______
8.	encourage students to communicate openly and honestly with others regarding their values	______	______	______
9.	help students use rational thought and emotions to examine their feelings and values	______	______	______
10.	aid students in critically analyzing their values	______	______	______

How to Interpret the Self-evaluation

Agreeing or being not sure about the following items indicates that you either tend to favor the following values education approaches or have not completely developed your philosophy regarding these approaches.

I, 3, 5, 8, and 9	- Clarification
2 and 6	- Inculcation
3 and 8	- Action Learning
4 and 10	- Analysis
5 and 7	- Moral Development

If you disagreed with all 10 statements, you either do not favor these values education approaches or you do not want to incorporate values education into your curriculum.

Some Values Education Research

Some interesting research findings indicate a lack of values education may be producing a society of individuals incapable of determining certain acts as being unethical. For example, Kurt and Gladys Lang at New York University and Amitai Etzioni at Columbia University gave a similar questionnaire dealing with the ability to rank unethical behavior to some of their undergraduate students.[2] The students could identify a behavior, such as one person stealing from another, as unethical. However, when the behavior dealt with an institution, such as one person cheating on income taxes, they were generally unable to recognize this act as being equally wrong. The social implications of these data could lead in the future to some troubling scenarios for our society.

Another informal study in this area tends to reveal why some students are incapable of determining if certain acts are unethical. Iozzi and Lockley, at Rutgers University, have found that students majoring in the humanities tend to reason at higher moral levels than those in the sciences.[3] If we assume that Rutgers' students are similar to most students in the United States, one may infer that science is one discipline that has primarily concentrated on cognitive and psychomotor skills rather than on the affective domain.

Earlier in this book, it was mentioned that a person's values were formulated through daily experiences. It was stated that people continually modify their values by being placed in situations that tend to conflict with their existing behavior patterns. If we accept this basic assumption regarding values and valuing, it would seem logical that an important function of education should be to give its students time to examine their beliefs and to sort out what it is they value as good or bad. By allowing students to make decisions concerning their values, educators can also provide experiences that will foster logical reasoning and critical thinking.

Why Some Educators Favor Moral Development and Clarification

Lawrence Kohlberg is a leading proponent of using the "moral dilemma" or moral development approach with students. Through this technique, students are not inculcated with certain values but are able to actively examine their own values and those of their peers by being confronted with a situation like this:

"What Would You Do?"

A genetic counselor's services have been sought by a couple who have a child with a serious genetic defect. The parents want to know their chances of having another child with the same defect. After some examination, the counselor finds that the genetic defect is not present in the husband's genes. Since both mother and father must carry the genetic information which will cause the defect in their offspring, the counselor knows that the child with the genetic defect must have been fathered by someone other than the husband.

Should the counselor divulge this information to the husband? (This particular example may not be appropriate for use in your classroom. Discretion must be exercised as to the type of moral dilemmas and/or moral issues that you present in your instruction.)

Kohlberg and others claim that by presenting moral dilemmas in classroom instruction, they have assisted students in raising their moral reasoning by one level based on Kohlberg's model for moral development.[4]

Kohlberg also believes that moral development is influenced by a person's religious background. However, he feels that it is possible to assist moral growth independent of religious education.[5]

Advocates of clarification appear to generally hold the same anti-inculcation philosophy as those favoring moral dilemmas. The difference is that they construct their activities with the main purpose of assisting students in developing the "valuing process."[6]

An example of this approach is as follows:

"Electricity and You"

List ten electrical appliances that you or your family use at home.

1.	6.
2.	7.
3.	8.
4.	9.
5.	10.

Circle those you feel you or your family could do without.

1. How many appliances on your list are essential for you and your family?
2. Make an attempt for one week to live without the electrical items you circled on your list. Keep a record of your progress for each day. Time will be allotted for you to share your list with others in class.

According to Sidney B. Simon, a leader of the clarification approach, students exposed to clarification techniques are able to critically analyze problems better than those not exposed to clarification activities.[7] He also contends that the students who have been introduced to the clarification approach have demonstrated the ability to follow through on their decisions. In other words, clarification has been shown to be an effective tool in developing better decision-makers.

Moral development, clarification, analysis, and action learning are all similar philosophically. By using moral development and clarification activities, students may have the opportunity to analyze their values and act in accordance with their beliefs.

We feel that allowing students to examine their own beliefs is far more pragmatic than attempting to instill in them someone else's values. For example, schools may attempt to teach that drinking alcohol at an illegal age is wrong. However, there is sufficient evidence to demonstrate that this will do very little to stop the underage consumption of alcohol. As one study indicates, peer pressure has the greatest influence on the alcohol drinking behavior of adolescents.[8]

When students are not in school, they formulate their own values on the experiences of their own environment, which includes the interaction they encounter with their peers. Giving them this opportunity in school may add tremendous relevance to their education. Therefore, we advocate all of the values education approaches, listed earlier, except inculcation. The activities in the last portion of this book reflect the philosophy of these four approaches.

Some Do's and Don'ts Regarding Values Education

The following are some suggestions about the use of the four values education approaches discussed in this chapter. Some examples have been included along with some specific suggestions for the purpose of clarity.

1. Create a classroom environment of trust. If your students feel they can respond freely and openly, they will tend to answer more honestly. (This type of environment does not occur overnight. Be patient!)

   Example: If a student professes to believe in racial discrimination, even though this may not agree with your beliefs, do not inform the student that his convictions are wrong. You can ask for other comments on this subject, but again be careful not to force your beliefs on the students. With differing views on a specific subject, students may be confronted with logical inconsistencies in their thought patterns. It is the intention of the four values education approaches (moral development, clarification, action learning, and analysis) to allow students to freely consider their own beliefs and to freely make choices on the basis of these considerations.

2. Make an all-out effort not to judge the responses of your students. It may be helpful at times to clarify their answers and comments, but refrain from offering your opinion unless you are asked for it.

   Example: A student may say that he feels that abortion is wrong. You might respond to the student by saying, "Why do you feel this technique is wrong?" Or, you may say, "Do you feel abortion is wrong in all circumstances?"

3. If you feel you must give your opinion on an issue, be sure you make it very clear that "this is my opinion."

   Example—Student: "What do you think, Miss Jones?" Miss Jones: "If you want to know my opinion, I believe that it is impossible to pass general legislation to deal with certain issues. Therefore, people must decide for themselves whether they want to be kept alive by a machine or not."

4. Be a good listener. In order to encourage student responses, you must be willing to listen carefully to all of your students.

5. Develop activities and experiences that are of personal and social concern to your students. If a student feels the subject discussed is very personal, respect his/her right not to respond.

6. Use divergent questioning whenever possible. These questions allow for open-ended responses, thereby encouraging a wider variety of student interaction.

   Example: What do you think are some of the ethical problems that a totally implantible artificial heart (TIAH) may pose to our society? This type of question allows for many possible responses and can be contrasted with the following convergent question that tends to limit a person's answer. What does Dr. Christian Barnard say about the effectiveness of the TIAH?

7. If the values activity requires student interaction, allow enough time for them to share their views.

8. Attempt to foster the type of environment in which your students can affirm their convictions without belittling the views of others.

   Example: If a student responds to one of his classmates, "That was really a stupid answer!", it is important to point out that you want your classroom environment to operate on the basis of mutual respect. Therefore, if you want someone to respect your opinions you must also respect their opinions.

9. Develop the distinction between respect for an opinion (acceptance of a person's right to hold a belief or opinion) and agreement with an opinion. While teachers need to respect student opinions, they certainly do not have to agree with them.

10. A student may perceive that his personal convictions are being attacked by either you or his peers even though you are non-judgmental of all comments or views. If this does occur, create a situation for the student to explain his feelings regarding the topic that caused the personal conflict. This may help the student recognize that his attitudes were so strong about a specific issue that any contradiction of his logic may have been misconstrued as a personal attack.

REFERENCES

[1]Etzioni, Amitai, "Should Schools Add 'Moral Education' to Their Curricula?" Minneapolis Tribune, October 17, 1976.

[2]Ibid.

[3]Piburn, Michael D., "Values Education: Insights From the Social Sciences," The Science Teacher, 43(3) 18-20, March 1976.

[4]Blatt, M. M., and Kohlberg, L., "The Effects of Classroom Moral Discussion Upon Children's Level of Moral Judgment," Journal of Moral Education, 4(2): 129-161, 1975.

[5]Mattox, Beverly A., Getting It Together, Pennant Press, San Diego, California, 1975, p. 20.

[6]The valuing process was discussed in Chapter 2. Its three main categories include: choosing, prizing, and acting.

[7]Simon, Sidney B., et al., Values Clarification, Hart Publishing Co., Inc., New York, 1972.

[8]Riester, Albert E., and Zucker, Roberta, "Adolescent Social Structures and Drinking Behavior," Personnel and Guidance Journal, 47:304-312, 1968.

4

SO WHAT'S ALL THIS HAVE TO DO WITH TEACHING SCIENCE?

Is Science Valueless?

Science can be defined as a systematic method of gaining knowledge about the world, the universe, and ourselves. This systematic method is characterized by various processes, some of which include: observation, inference, communication, predicting, hypothesizing, and controlling variables during an experiment.

However, we feel that science also has both nonmoral and moral values components. To explain this position, let's take a look at a problem in Michigan that was identified in 1973.[1] Farmers were complaining that their cows were exhibiting strange symptoms, including refusal to eat which eventually lead to death. A local veterinarian became extremely alarmed and sent some fat samples, from cows that had died from the bizarre illness, to Michigan State University for analysis. Michigan State concluded that the animals died from starvation. The veterinarian knew the animals had starved because of their refusal to eat. However, he still wanted to know the reason why they would not eat. In an attempt to have this question answered, he sent more fat samples to a private research institute in Wisconsin and to Purdue University. The results from these institutions indicated that a fire proofing chemical called PBB was present in high amounts in the fat tissue.

About the same time, a scientist at Michigan University arrived at the same conclusion. The veterinarian and the scientist felt compelled to make this information available to the public. However, they were subjected to harassment when they attempted to divulge the PBB information. It appeared that state agriculture officials had made a slipup. A chemical company who produced PBB and a nutrient for cattle feed made a shipping error, causing PBB to be added to cattle feed instead of the nutrient, thus resulting in the accidental poisoning of hundreds of cattle. The state agencies in charge of monitoring cattle feed production had failed to catch this error. If this mistake was publicly announced, any farmers who lost cattle from the PBB poisoning would be entitled to state reimbursement for their losses. In addition, any person consuming any products from the contaminated cattle would also be subject to secondary PBB poisoning. Even though this situation was extremely serious, some state officials attempted to stop the veterinarian and the scientist from reporting the PBB poisoning.

If we analyze the Michigan situation, it is apparent that the veterinarian cherished the lives of the animals and believed that his clients were not responsible for their deaths. Both the veterinarian and the scientist demonstrated a high value in the scientific method to assist them in solving their problem. These three examples would fall into the category of nonmoral values. Besides these, they both had concern for the health of any humans involved in a secondary poisoning from the PBB and felt a moral responsibility to reveal the truth concerning the poisoning, regardless of any physical or verbal harassment.

Because the scientific enterprise places a moral responsibility on its practitioners to communicate all of the data they receive and because scientists are motivated to undertake certain areas of investigation by the personal values they hold, we feel science cannot be described as a valueless process. It would be more accurate to think of science as an enterprise comprising knowledge, processes, and values. Therefore, we advocate presenting all three of these components of science in the classroom.

Where Is Science Education Regarding Values Education?

During the last two decades, science education in the United States has supported the teaching of science facts, concepts, and processes. The "new science curricula" of the late 1950s and early 1960s did not contain any values education. For example, the Biological Sciences Curriculum Study (BSCS) was originally funded by the National Science Foundation (NSF). NSF made it very clear to BSCS that in order to receive financial support from the Foundation science values issues must be omitted from their programs.[2]

In the late 1960s and early 1970s, values education was incorporated into a few science curriculum programs. These programs were: Technology, People and Environment,[3] Investigating Your Environment,[4] Human Science Program,[5] and some BSCS supplemental audiovisual materials. Around the same time that these science materials were developed, some science educators expressed a need for values education in various science education journals. However, there still appears to be a lack of values education in most science curriculum programs, science textbooks, and science instruction in the United States.

Science Instruction That Presents Knowledge, Processes, and Values

We recommend presenting science knowledge (facts and concepts), processes, and values in your instruction. In most cases, this recommendation is suggesting that you merely extend your present science material to touch the personal lives of your students. For instance, a topic that is common to most biology textbooks is genetics. Most books present the material in the following way:

I. Science Knowledge (Facts and Concepts)

Basic Mendelian principles are presented by using pea plants, fruit flies, etc., as examples of how these principles apply to living things. Hypothetical situations are contrived showing the expected offspring of various plants that have cross-pollinated or animals that have mated.

II. Science Processes

From these situations, students learn how to predict what offspring can be expected in specific plants and animals, regarding certain inheritable traits. Some science programs encourage the students to design their own experiments, using fruit flies or other organisms, to investigate some of Mendel's principles. The students may also be asked to make predictions regarding their possible experimental data.

Most science teachers agree that these science facts, concepts, and processes are essential to the basic understanding of the field of genetics. But, at this point students may not be able to transfer this information into something that could be potentially valuable to them now and in the future. Therefore, we propose the inclusion of topics and issues that relate directly to your students' lives.

III. Values

For example, if students studying genetics are introduced to various techniques, such as genetic counseling or screening, they may have a greater chance of recognizing how the basic Mendelian principles that they have been investigating can assist them in gaining valuable information about their family and their future. If these techniques and others are pursued further, the students can also be confronted with some of the possible implications that this field of biology poses to society. In other words, students would be able to consider some of the values issues that are present in this area of science and would simultaneously be given the opportunity to recognize how science is applicable to their lives. An example of an activity that deals with the values level of genetics is provided below.

Reprinted from THE AMERICAN BIOLOGY TEACHER, Volume 37, No. 3, March 1975; courtesy of the National Association of Biology Teachers, Inc., 11250 Roger Bacon Drive, Reston, Virginia 22090.

Genetic Counseling

Genetic counseling provides and interprets medical information based on expanding knowledge of human genetics. Its major goal is to prevent the occurrence of birth defects that are due to genetic inheritance.

Although individual birth defects may seem to be relatively infrequent—occurring in about 7 percent of all births—each year about 250,000 American babies are born with physical or mental defects of varying severity.[6]

Using the basic laws of genetics, a genetic counselor can predict the probability of recurrence of a given abnormality in the same family. For example, Tay-Sachs disease is inherited by a recessive gene. A child afflicted with this disease begins to show signs of central nervous system degeneration at about six months. These symptoms continue to increase, usually ending with death before the child's fourth birthday.[7]

If both parents are carriers of Tay-Sachs, they will have a 25 percent chance of having a Tay-Sachs child with each pregnancy. The following diagram illustrates this:

Symbols of Genes: T = Normal Gene
t = Tay-Sachs Gene

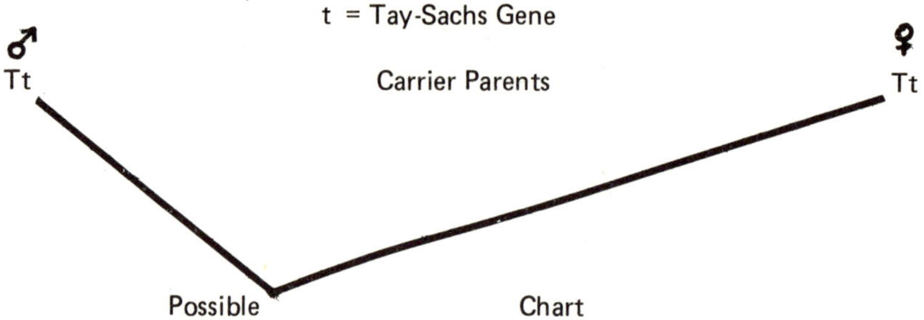

Possible Offspring:

	T	t
T	TT	Tt
t	Tt	tt

Chart Interpretation:

TT 25% both normal genes
Tt 50% carriers
tt 25% Tay-Sachs

Please respond to the following statements and questions. Time will be allotted to discuss your responses.

1. Genetic counseling will benefit mankind.

Agree -----
No opinion -----
Disagree -----

2. I feel that, at some time, I would consider using the services of a genetic counselor.

Agree ------
No opinion ------
Disagree ------

3. Genetic Counseling should be a free service to all members of our society.

Agree ------
No opinion ------
Disagree ------

4. Students in high school should be made aware that genetic counseling exists.

Agree ------
No opinion ------
Disagree ------

5. The government should provide adequate publicity of genetic counseling on the TV and radio, so more adults would be aware of this service.

Agree ------
No opinion ------
Disagree ------

6. What is the difference between genetic counseling and genetic screening?

7. Should a genetic counselor have the legal right to prescribe medical treatment (i.e., sterilization, abortion) for a couple known to be carriers of a genetic disease? Explain your answer.

8. Do you feel the government should require couples that have a child with a genetic disease and are intending to have more children to seek the services of a genetic counselor? Explain your answer.

Other activities designed to incorporate values education into your science instruction comprise the second half of this book. They may be easily applied to most existing science courses.

REFERENCES

[1]This example was adapted from the WETA, Washington, D,C., TV presentation, "The Poisoning of Michigan."

[2]Information received from a conversation with BSCS administrators in Boulder, Colorado, 1973.

[3]Technology People and Environment, Learning Realities, Inc., 10 East 40th St., New York, New York 10016.

[4]Investigating Your Environment, Addison-Wesley Publishing Co., Inc., Reading, Massachusetts 01867.

[5]Human Science Program, is being developed by BSCS and will be completed in 1981.

[6]Genetic Counseling, booklet printed by the National Foundation for the March of Dimes.

[7]Stine, Gerald James, Biosocial Genetics: Human Heredity and Social Issues, Macmillan Publishing Co., Inc., New York, New York, 1977.

5

VALUES, SCIENCE, AND CONTROVERSIAL ISSUES

A Case Study

An unprecedented event in modern science took place in July, 1976, when the Cambridge, Massachusetts, City Council legally restrained Harvard University from conducting recombinant DNA research until the City Council could decide whether or not to allow this research within Cambridge's city limits.[1] This event was historic because it marked the first time that a governmental division in this country had attempted to restrict scientific research by legal means. The action grew partly out of the spirit of the times in America where people look at science with a more critical eye than in times past, and partly out of the nature of the research itself.

The research centers on an important chemical substance, DNA. DNA (deoxyribonucleic acid) is the material of which chromosomes are comprised and which contains the chemical information which determines all the biological characteristics of organisms. Recombinant DNA research evolved as the result of biochemical techniques which allowed researchers to break apart small pieces of DNA and then reattach them to other pieces of DNA.[2] When this technique is accomplished, and the new (recombined) DNA is re-introduced into the cells of living organisms, the organisms reproduce and the new recombined DNA reproduces with them. Often, the new DNA will cause unique changes in the organisms, or in the products they manufacture. Recently, such research has resulted in the ability to "teach" a strain of bacteria (by altering its DNA) to produce the substance somatostatin.[3] Somatostatin is believed to have potential therapeutic value in acromegaly, acute pancreatitis, and insulin-dependent diabetes. Thus, this new technique is already offering great potential for the promise of synthesis of many drugs and substances that are important to people.

Despite the positive potentials for recombinant DNA research in the areas of drug and biochemical synthesis, and a quantum leap forward in understanding human genetics, genetic disease, and cellular physiology, there are also some potentially serious problems with the research. Both scientists and laymen have criticized the new area of research by pointing out that this technique may possibly produce real hazards to humans and the natural environment.

The common experimental organism *Escherichia coli (E. coli)* is the bacterium normally found in the gut of many animals and is an inhabitant of the human gut as well. The critics fear that a laboratory accident could allow a genetically altered E. coli, perhaps with pathogenic characteristics, to escape into the environment, and cause an epidemic disease for which we might have no cure. They are also concerned that "new" organisms created by this technique might have serious ecological impacts in the natural environment. In addition, they fear the potential misuse of these techniques in developing biological weapons, or in other ways that could endanger people.

As the techniques developed between 1972 and 1974, scientists working in this research area began to disagree among themselves as to the dangers and risks involved in the research. Some felt that the research offered so many potential benefits, both in terms of the advance of scientific knowledge and the direct application of the research to medical problems, that it should proceed as rapidly as possible. Others felt that the potential problems merited serious attention and that careful analysis of these risks should precede the research.

This controversy grew to the extent that a conference at Asilomar, California, was convened on the subject in February of 1975.[4] At the conference, the scientists eventually agreed to a moratorium on recombinant DNA research until a strain of E. coli which could not survive outside of specialized laboratory conditions could be perfected. In addition to this control, the group of researchers began to participate with the National Institute of Health (NIH, the Federal governmental funding agency through which most recombinant DNA research proposals would be funded), to develop specific guidelines for the conduct of such research.[5] These guidelines set up very specific laboratory procedures which must be followed for a variety of levels of research if the NIH would fund the research.

It was following the Asilomar conference and the development of the NIH guidelines that the controversy between Harvard University and the community began. The mayor of Cambridge brought the issue before the city council and a committee was appointed to examine the research and the guidelines. This committee finally recommended approval of the proposed research, but stipulated several additional safety regulations for the research in addition to the NIH guidelines. In effect, their action proved that informed citizens not only had the right to participate in examining such research, but that they could do so effectively. While not everyone at Harvard or in Cambridge may have agreed with their recommendations, public policy related to scientific research had been formulated in a manner in which the people could participate. This situation demonstrates that controversial issues relating to science cannot only be understood intelligently by people other than scientists, but that such controversy can be resolved.

It seems quite likely that such controversies which relate to science and the citizen will continue to develop in the future, perhaps at an increasing rate. This means that our students will have an increasing opportunity for becoming involved in such controversies, and that, therefore, they can profit from analysis of such issues and activities in their formal education. A working knowledge of how to investigate such issues and how to participate effectively in public science policy decision-making will be an increasingly valuable asset to young people.

Controversial issues are often avoided by classroom teachers because they either feel that such topics should not be discussed in the school or because they feel unprepared to deal in an instructionally effective way with these issues.

Just what is a controversial issue? The word controversy has two definitions in Webster's New World Dictionary.[6] One definition is: "A quarrel or dispute." The other is: "A discussion of a question in which opposing opinions clash." While acknowledging that people often approach controversy by quarreling, we would stress the notion that the second definition is the appropriate approach for facilitating an instructionally effective classroom session around a controversial issue. Students should be helped to learn how to engage in the process, take into consideration rational arguments which are carried by both sides, and recognize the critical roles which differing values play in the issue. We contend that this is a very important component of science education in contemporary society.

When we discuss scientific controversies, it is important to recognize the fact that there are values held by scientists which influence their positions. In this next section, we will explore some of these values and attempt to show how they may conflict with other values.

Some Values of Science

In 1966, the Educational Policies Commission of the National Education Association published a document entitled Education and the Spirit of Science.[7] The Commission felt that schools should promote the understanding of what they called the values of science because they saw those values as growing out of a view of human dignity and not out of tradition. They believed that the values they endorsed were characteristic of rational thought in all areas of human endeavor, not only in science, but that they were perhaps most easily identified in scientific activity. The values they outlined were:

1. Longing to know and understand
2. Questioning of all things
3. Search for data and their meaning
4. Demand for verification
5. Respect for logic
6. Consideration of premises
7. Consideration of consequences

While aspects of the presumptions of the statement, and other inadequacies have been voiced by others,[8] we believe these "values" (perhaps better defined by David Hawkins as "virtues") do offer a useful model for analysis of controversial issues as instructional vehicles.

Let's re-examine the recombinant DNA issue cited earlier in this chapter in light of the Educational Policies Commission's statement. Certainly, the longing to know and understand was a prime motivating factor in the development of the techniques leading to the recombinant DNA research and to the concerns about it. Also involved in the minds of the researchers was a questioning of all things. This second "virtue" was also held by critics of the research who would not, in blind faith, accept the claim that the research posed no threats.

The Cambridge City Council's committee spent months searching available data for their meaning, and demanded verification of claims made by proponents and opponents of the research. In the context of their deliberations, the committee attempted to operate rationally (respecting logic) rather than politically or emotionally in dealing with this issue. They carefully considered the premises of the sides of the issue, and were exceptionally sensitive to the consequences of their final decision to allow the research to proceed providing the NIH guidelines and certain additional stipulations they prescribed were followed. The committee operated in a careful, rational manner, exhibiting the virtues which the scientists themselves might be expected to display in their research. Their actions indicate that they valued this method of behavior in decision-making and problem solving. Most people would agree that approaching controversy with such behavior is more desirable than non-rational methods, primarily because the virtues which characterize this method treat human beings as equals and attack the issues and not other people. The method demands that all views be evaluated by the same objective rules. In other words, these virtues can aid people in the "valuing process".

The scientific community has historically espoused this rational approach characterized by these seven virtues as a means of valuing and evaluating its own work. Although there are cases in the history of science in which the scientific community has failed to live up to this standard, the commitment has persevered and persists today. The virtues and method have generally been successful when put to the test of use.

To return to the definition of values we developed earlier in Chapter 1, the value placed upon using these virtues in problem solving is a moral value. That is, it is an assessment of what is considered to be good actions and motives with respect to others (i.e., the ideas of others). This moral value, of course, can conflict with other values. Common folk wisdom often moralizes with: "Curiosity killed the cat!" or, "What's good for General Motors is good for the U.S.A.!" or similar statements of values which stand in opposition to the virtues of science. People are often swayed by the opinions, arguments, and information which support their own particular point of view on a given issue. Often, people make decisions without examining the consequences of their decisions. In other circumstances people adopt the position of their particular peer group without questioning the logic of the position. And, unfortunately, opinions are often developed reactively. That is, people take a position in opposition to that held by a group of other people whom they regard as inferior in some way or other. Whenever people value an ideology, their prejudicial beliefs, the acquisition of material goods, or status more than they value knowing the truth as completely as they can, their values are likely to come into conflict with a valuing of the method characterized as the "spirit of science."

Unfortunately those values which stand in opposition to the spirit of science often serve to polarize people's decisions and further estrange them from one another. We believe that the method and virtues characterized by the spirit of science have great potential for resolving controversy, and we feel that you can facilitate the development of the spirit of science in your own classroom. Developing this approach to conflict resolution must, however, grow out of the experiences of your students and not out of inculcation. Inculcation is a means which is inconsistent with the ends of the spirit of science. Rather, we suggest that it grow out of the context of real experiences in which your students discuss and analyze controversial issues which are familiar to them.

Making Decisions

We live in a culture which is competent to deal with various forms of technology. Electronic calculators by the thousands are bought and used. People are not intimidated by common sophisticated technical devices, and prove to be quite capable of operating such equipment with minimal experience. We are devoted to technology, capable of using and consuming technological materials, but we are also a culture which is not well informed of the basic scientific methodologies and ways of thinking which form the framework upon which our technology is built.

Science education is today becoming more directed at public scientific literacy than ever before. Part of this literacy ought to include an understanding of the values system which characterizes work in science. Previously, we claimed that the spirit of science had potential beyond the realm of science; however, it is also obviously important in science itself.

As you consider using controversial issues in your classroom, what kinds of criteria can you use to evaluate the quality of the learning experience? In short, what are your objectives in using these types of issues, and how can you make some sort of evaluation of whether or not they are being accomplished through your methods?

We think that the list of virtues we identified as values of science earlier in this chapter can be one means of judging the effectiveness of your class discussions of controversial issues. As students interact on controversial issues, you should be able to assess, primarily through observation, but also in other ways, the extent to which your students are exemplifying and exhibiting the seven virtues described by the Educational Policies Commission. Generally, these types of objectives would be classified as part of the affective domain. That is, they relate to the attitudes and values of the students. While we feel that evaluation in this domain is quite difficult, we believe, as do other science educators, that it can be done with reasonable effectiveness through careful observation.[9] We also think that objectives, whether they are cognitive or affective should never be "hidden agenda"! We believe that your students should know what your objectives are when you set up instructional situations in your classroom. The seven "values" proposed by the Educational Policies Commision could serve as a good set of instructional objectives for your discussions of controversial issues.

This does not mean that you must necessarily restrict all discussions to the development and accomplishment of those objectives. Students come into discussions of controversies with their own views, beliefs, and opinions. They come by these honestly, however inadequate they may appear to be to you. They should not be criticized by you for their views. During the classroom interaction, they will have the opportunity to share their views, compare them with those of others. They will have the opportunity to listen to new information, discover new premises, analyze consequences, and have their logic criticized by their peers. This type of social interaction occurring in the context of discussion of a controversial issue is a critical and highly potent learning environment, and does not require as much control or structure as other learning situations. Be careful about intervening in these discussions. Often, your concerns in a student discussion will be dealt with by the students without your saying anything, and at a more opportune time and in a more effective way!

It is also important to recognize that the spirit of science takes time to develop. These behaviors will not appear miraculously at the close of the first discussion of a controversial issue. They are not facts or principles, but rather are concepts and ways of coming at facts and principles. Therefore, a number of experiences in working with these values are necessary over the course of a school year, before you might expect to see behaviors in your students which are indicative of their having attained these objectives.

Finally, it is essential to recognize the limitations of any method for resolving controversial issues. Your students will not always reach consensus on issues, nor should you expect consensus. Often the various positions in a controversy grow out of different basic premises. At this point, controversies become exceedingly difficult to resolve. Nevertheless, this can also be used as an important learning opportunity. To be aware of the fact that different premises can, and usually do, result in different conclusions, is quite essential if students are to become more understanding and accepting of the views, opinions, and beliefs of others.

REFERENCES

[1]Cullitan, Barbara J., "Recombinant DNA: Cambridge City Council Votes Moratorium," Science, 193 (4250): 300-301, July 23, 1976.

[2]Cohen, Stanley D., "The Manipulation of Genes," Scientific American, 1:24-33, July, 1975.

[3]Itakura, L., et al., "Expression in *Escherichia coli* of a Chemically Synthesized Gene for the Hormone Somatostatin," Science, 192 (re21): 1056-1063, December 9, 1977.

[4]Berg, Paul, et al., "Asilomar Conference on Recombinant DNA Molecules," Science, 188 (4192): 991-994, June 6, 1975.

[5]Op. cit.

[6]Webster's New World Dictionary, World Publishing Company, Cleveland & New York, 1958, p. 322.

[7]Education and the Spirit of Science, Educational Policies Commission, National Education Association, 1966.

[8]Hawkins, David, "Education and the Spirit of Science: Critique of a Statement," The Science Teacher, September, 1966, pp. 18-20.

[9]Sund, Robert B., & Picard, A. J., Behavioral Objectives, Charles E. Merrill Pub. Co., Columbus, Ohio, 1972.

6

CAUSES OF CONTROVERSY: SOME PERSPECTIVES

The values held by people grow out of their experiences and shape their behaviors. In other words, values are learned and can be inferred from observation of people's behavior. The rate at which automobiles of certain types are bought, the kinds of records which sell millions of copies, and the success of quick food franchises are conditions which might allow us to infer some of the values of the general population of the U.S. The sum total of values and behavior of a people are often characterized as their "life-style." While it is difficult to claim that a particular life-style is either good or bad, some people have claimed that there are components of life-styles in this country which should be changed or serious consequences will follow. Such claims usually result in the advocate of change being ignored, but occasionally there is a public reaction and a controversy will result. In this chapter, we will attempt to get some perspective on the causes of controversy so that as you deal with controversial issues in your classes, you will have more insight into the origins and nature of the problems you choose.

A Case Study

A great public controversy developed in rural Minnesota in the winter of 1978. Farmers demonstrated and were arrested. Construction workers were intimidated, and the Governor of Minnesota considered calling in the National Guard. The cause of this controversy was a high voltage direct current power line which was being built across farm land from North Dakota to supply the energy requirements of the Minneapolis-St. Paul area. The state of Minnesota had granted the power utility the right to construct a modern direct current line over the farmland. The farmers contended that not only were their rights to their property being violated, but that the power line posed a threat to their safety and health because of the high degree of radiation which was predicted to come from the power line once it became operational. While recognizing the need for energy of the inhabitants of Minneapolis-St. Paul, the farmers perceived the power line as a threat to their property and their health. They claimed that the State's own agency, the Department of Natural Resources, had prevented the building of the line on state-owned wildlife land because of concern over environmental dangers which it may have caused. If, they claimed, the state was so concerned about the possible threats of the line to wildlife, how could they feel justified in putting it directly over farmland and the heads of farmers?

In this case, a variety of values interacted to cause the controversy. Economics, politics, geography, and the cultural norms of the region brought about a serious conflict. Interestingly, one of the proposed solutions to the controversy was to institute a "science court" which would examine all the evidence and research on potential environmental effects of high voltage direct current. This court would then rule on whether or not to allow the line's construction through the farmland. While this proposal has not been successful in being adopted, the situation indicates that scientists and scientific knowledge may have real importance in areas of controversy. Whenever issues involving life-styles, human safety, or economic concerns are involved, controversy can logically be expected to develop. Being scientifically literate will be important in dealing with such controversy, but understanding how human values develop and are influenced is also critical in such problem solving situations.

Human beings have been investigating moral values for thousands of years. As we mentioned in the second chapter, values can be identified as being either non-moral (i.e., evaluating what is good or bad with respect to objects) or moral (assessing good or bad actions and/or motives with respect to others). Out of much of the work which has been done in the area of moral values, or ethics, it is possible to develop a view of three major "systems" involving ethical beliefs. Although these systems do not adequately describe every unique ethical system which has been devised by people over the years, the three major groups are useful in understanding the bases upon which ethical thought lies. The three groups we will examine are identified as "Principle," "Consequence," and "Context" ethics.

Principle Ethics

People who ascribe to the belief system known as "principle ethics" generally believe that there are some universal rules or norms related to how people treat and interact with one another. This belief system was championed by people such as Immanuel Kant and has wide acceptance throughout the world.[1] Kant developed what he called the "categorical imperative" which, in common language, says that human conduct with respect to others should be judged on the basis of whether or not it could serve as a universal rule for all people. Kant also advocated that people always be viewed as "ends" and never be treated as "means" to other ends. Someone operating with a system of principle ethics would contend that there are certain universal principles which are accepted by all people, such as the value of human life, the value of truthfulness, etc., which serve as the basis for ethical conduct.

An example of principle ethics could come from the following situation:

> Rachel and Jane are good friends. Jane is dating Frank and is planning to marry him upon their graduation from high school. Last night, Rachel saw Frank out with Joan, a new girl in town. Frank and Joan were very friendly and seemed to be quite engrossed in one another. Rachel knows that telling her friend Jane of Frank's unfaithfulness will be quite painful for her, but Rachel believes it is important to tell the truth, "no matter what happens." Rachel's motivation for telling Jane about Frank is based upon her use of principle ethics; namely, a rule which says that honesty and truth-telling regardless of the consequences is essential in her behavior as a human being.

Another example of the use of principle ethics might be the view of the person who is opposed to any abortion for any reason on the basis that "all human life is sacred and must be protected at all costs." This principle guides the ethical conduct of that person, and is another example of principle ethics.

Consequence Ethics

Thinkers in the 1800's such as John Stuart Mill and Jeremy Bentham were advocates of another system which we shall call "consequence ethics."[2] They claimed that pain and pleasure were the primary motivating factors of human beings and, therefore, a kind of "moral calculus" could be derived. They believed that every ethical problem could be analyzed and the degree of "goodness" or "badness" of a particular action could be measured on the basis of the extent to which it provided pleasure and avoided pain. In other words, they contended that the consequences or results of actions determined their morality. The extent to which an act had consequences which increased pleasure and/or decreased pain would determine whether or not it was a moral act. The notion of "the greatest good for the greatest number" has evolved in contemporary thought from this view and is widely used in human decision-making.

Using the situation we developed earlier with Jane, Rachel, Frank, and Joan, let's examine how consequence ethics might apply. If Rachel is motivated by consequence ethics, she might begin by asking herself, "What should I do that will cause the least pain and bring the greatest pleasure to everyone involved?" She may then reason that to tell Jane of Frank and Joan might cause a great deal of pain. Frank's brief interlude with Joan may have been nothing serious and Rachel would have intervened in a situation only to bring trouble to everyone. Frank and Jane's relationship might be destroyed and Joan may acquire the reputation of being someone who breaks up relationships. Rachel may reason that Jane will eventually be "better off" without Frank and that Frank and Joan might be happier with each other.

Using consequence ethics, Rachel will have to "weigh the possible outcomes of her action," and her decision will be based upon this kind of "moral calculus."

To use the abortion issue, a person operating with consequence ethics might reason that an unwanted child might cause more pain to the mother than the abortion and hence be in favor of that kind of solution to the pregnancy.

Context Ethics

Context ethics developed in the 20th century primarily out of the thinking of such people as Jean-Paul Sartre and, more recently, Joseph Fletcher.[3] An outgrowth of existential philosophy, this belief system operates under the premise that the moral value of an act can be determined by whether or not a person has taken the action. In this system the action taken by the person in a situation determines whether or not the act is moral. While this sounds rather confusing, people who support context ethics propose that moral choices only grow out of given situations or contexts and that each situation must be analyzed and understood to see the morality of the actions taken. People such as Fletcher use the dictum: "Go and do the loving thing." Often referred to as "situation ethics," this view has caused controversy among people who study philosophy and ethics.

How would Rachel operate with context ethics in her dilemma? Rachel might reason that the situation is such that she should do nothing until such time as she has evidence that Frank and Joan are indeed serious about one another and that Frank is "two-timing" Jane. She may feel that the situation at present does not merit her intervention, but that it might change and then she would act. Rachel is in effect indicating that the context in which her particular problem arose demands silence at one time and action at another. She might also reason that the "loving thing to do" is to tell her friend Jane about her concerns about Frank and Joan, and that as long as Rachel's motives are pure (i.e., to do the loving thing) everything will turn out for the best. Both reasoning approaches would indicate the use of context ethics.

Someone examining the issue of abortion from the perspective of context ethics might react by saying, "Well, it all depends. In some cases abortion might be O.K. and in others it might be wrong; it would depend on the circumstances of the people involved." This person would be operating out of context ethics regarding all abortion.

You have now had an opportunity to examine the three ethical belief systems and some examples of how they might be used by people in ethical decision-making. Now, read through the situation below and the two hypothetical responses. After reading the responses, attempt to identify the ethical belief system or systems which the two respondents are using.

SITUATION NO. 1: There is a 66-year-old surgeon who serves 50,000 people in a remote area and he needs a kidney transplant. The donor would be his 22-year-old grandson who is a promising medical student. The grandson is the father of two young children with a third on the way. Statistically, it is known that, on the average, the donor of a kidney shortens his own life by five years. Since the "matching" of the kidneys in this particular situation is quite difficult, it is unlikely that a suitable donor other than the grandson will be found.

RESPONSE NO. 1: "I think the transplant should be done. The morally right response is always to save a life. You can't play the statistics game with the grandson. An average is just that; the grandson may live an entirely normal life with no ill effects, and the grandfather can have a number of productive life-saving years ahead of him as the result."

RESPONSE NO. 2: "If you weigh the pros and cons of this situation, I think you would have to come down on the side opposing the transplant. The grandfather is nearing the end of his career and the grandson has the potential for doing highly significant work which may benefit millions of people, not to mention the responsibilities he has to his own family."

Did you have any difficulty identifying the ethical belief systems operating in Response Number 1 or Number 2? If you had some difficulty, check back over the systems and the examples outlined earlier and then proceed.

The first response to Situation Number 1 indicates that the person is concerned about human life, and saving one whenever possible. This person is not weighing consequences or equivocating about situations; there is a moral principle involved and it is being applied: "Save lives regardless." This person is using principle ethics. The second response to the situation shows that the person is also concerned about human life. In this instance, however, it seems that the person has decided that more good can result from not performing the operation than from making the transplant. This person is using consequence ethics.

If you had no problems with correctly identifying the systems used in the first situation, great! Go on, now, read Situation Number 2 and the two responses to it, and attempt to analyze the ethical systems which are motivating those responses.

SITUATION NUMBER 2: A genetic counselor's services have been sought by a couple with a child who has a serious genetic defect. The parents of this child want to know what their chances are of having another child with this defect. After some examination, the counselor finds that the potential for this genetic defect is not present in the husband's genes. Since mother and father must both carry the genetic information which will cause the defect in their offspring, the counselor knows that the child with the genetic defect must have been fathered by someone other than the woman's husband. Should the counselor divulge this information to the husband?

RESPONSE NUMBER 1: "I think the counselor will have to get more information on the situation regarding the couple. I think his professional responsibility is to tell the truth, but it has to be done in such a way as to not destroy the relationship. If he finds that his information may endanger the relationship, he should do the thing which will harm the couple the least, and not tell the truth about his findings."

RESPONSE NUMBER 2: "If I were in that situation, I'd be forced to tell the husband. As a professional, I'd be bound to maintain my integrity and honesty, regardless of the outcome. The fact that the couple seems to be interested in having more children seems to be an indication that their relationship is good, but under any circumstances, I think that knowing the truth about the situation is the only basis for any relationship."

In Response Number 1, the person begins by mentioning his or her regard for the principle of truth-telling, but then goes on to conclude that the proper decision should be based partly upon another principle; namely, that human relationships should not knowingly be destroyed or jeopardized. This person's statement that the action which will harm the couple the least might be alternatively interpreted as being either consequence ethics or context ethics. We believe this is a complex response, but that it probably shows context ethics to be most operative in this person's thinking.

The second response seems to be a more clear-cut one. The person apparently values the principles of integrity, professional honesty and truth-telling and justifies the use of principle ethics by stating that the truth is the only basis for a relationship.

We hope that these example situations have given you a better grasp of the three ethical belief systems. Although it may appear that this type of analysis is difficult, we think you will find that you can acquire more skill in understanding the perspectives of your students, as you listen to their discussions, with this type of model. Also, it is important to recognize that this is only a model. It is not ethical behavior in its reality. This model, as others we have presented earlier, should be seen as a help to you in organizing your thinking about your students' moral, ethical, and value behaviors. Don't try to "force" their positions or views into one of the three systems. Rather, attempt to distinguish consistencies, inconsistencies, and the complicated combinations of these belief systems which people employ as they work through moral problems.

Whenever choices related to other people are made, ethical belief systems are employed. Certainly, people in real life situations use combinations and/or bits and pieces of all three systems to justify their actions. It would be unwise to suppose that personal consistency is an important criteria for assessing someone's morality. Nevertheless, our ethical belief systems, examined or unexamined, do influence how we make moral decisions. It follows then, that whenever people having differing ethical belief systems investigate a problem involving moral values, they may conflict or clash regarding the proper decision.

Each of the three ethical systems has its difficulties in certain situations and no single system can be rationally preferred over the other. Those who employ principle ethics often find themselves in situations in which two or more of their principles are in conflict. Those who advocate consequence ethics often cannot clearly identify the decision which will provide the greatest pleasure and the least pain. People adopting a position based upon situation, or context ethics, are often forced to conclude that the "loving" thing to do cannot be determined absolutely. As a result of the facts that people generally are unaware of their ethical belief systems, and that these systems all possess inadequacies, controversy over moral questions is practically assured.

When you are involved in classroom sessions that deal with questions which involve moral values, it is important that you can recognize the belief systems being used by your students and yourself. When belief systems conflict, the major premises used in the discussion are in conflict. Under such circumstances, "right" or "wrong" decisions are tied to the different premises and belief systems. Helping your students to understand that their premises and belief systems are in conflict is an important first step in understanding the nature of the problem. In classroom situations dealing with moral values, your goal should be one of developing an understanding of the differing premises which cause the controversy of values conflict and not one of resolving the controversy. If you can do this effectively, you will be encouraging rational analysis of problems and will be developing a spirit of tolerance among your students. To disagree with someone else is not "bad" provided you know why you are in disagreement!

Non-moral Values

There are many factors which influence our non-moral values. As we experience life, we adopt goals, adapt to our environmental surroundings, and make many decisions relative to our non-moral values. Life-styles, cultural norms, economics, politics, religion, and even geography will influence our decision-making. As you select topics and issues for classroom interaction, your knowledge of how these factors influence the valuing processes of your students is crucial in understanding how to choose and facilitate an effective learning session. Let's investigate some of these factors and analyze their influence on your students' values.

Economics

America is a mass of economic complexity. Probably the vast majority of problems issues and the decisions reached relative to them are influenced heavily by economic considerations. The economic activity in an area can, to a large extent, determine whether or not a particular issue or topic will be controversial. Many non-moral values are related to economics where the emphasis is on the value of materials and services. Often, non-moral values come into conflict because of economic considerations related to them.

As an example, let's consider a situation in which the Environmental Protection Agency (EPA) has investigated air pollution in a small community and has ruled that the single large industry in town is polluting the air and must either come into compliance with Federal clean air standards or shut down. Economically, the plant may not have a sufficient economic base to bring its operation up to Federal standards. If the plant shuts down, the economy of the town will be destroyed, people will be unable to find employment, and financial disaster for everyone in the town is imminent.While most people in the community probably value clean air, and would even support Federal efforts to maintain it, the EPA ruling becomes quite controversial in the community as it becomes apparent that economic hardship will result from the action. A classroom activity on air pollution in this town poses much more potential controversy than it would in other areas where the local economic problem does not exist. Similarly, the people in a town adjacent to a large military base may value cutbacks in government spending.

However, when the Defense Department decides to abandon the military base, the value of financial security quickly takes precedent over the prior value. Finally, a teacher in a health class may decide to teach a unit on the health hazards associated with smoking. While most people value health quite highly, this teacher quickly becomes embroiled in a local controversy because he is teaching in an agricultural area whose main product is cigarette tobacco.

All of these examples point out the extent to which economic considerations influence our non-moral values. It may be possible that some of the most significant problems in our society today are centered around conflicting non-moral values, and teachers who can facilitate constructive study of such problems are certain to be successful in educating their students for the future.

Cultural Norms

The standards or "norms" of a culture are taught to children throughout their lives. In adolescence societal norms are often challenged by the norms of the teenage subculture, but overall, the non-moral values of people are adjusted by experiences resulting from living in a culture which has developed certain ways of solving problems and acting. The so-called "work ethic" which seems to permeate most of American culture has been under fire for years. Yet, most Americans still value work, and react negatively to others who do not.

The American culture is also one which has come to value private transportation. The automobile has become such an integral part of our lives that life without it is unthinkable. While the cultural norm may dictate the "two car or three car family," there is a growing realization that this value is in conflict with other values such as wise use of resources, conservation of energy, clean air, and a safer life.

The values of the population as a whole, as characterized and inferred by their collective actions and behaviors, provides an excellent background against which students can thoughtfully examine their personal values. The social changes which typify contemporary life are bound to generate conflict and controversy. Areas of social change which relate to non-moral values can and should be distinguished from areas relating to moral values.

Gay rights, civil liberties, social welfare, and single parent families are among the examples of moral value issues which are in conflict with cultural norms. Many of the moral values of the culture are being questioned and examined today, in an effort to discern whether or not the reasons for their development were legitimate and if so, whether or not they are justifiable now. Your students may either accept or reject various cultural moral values, and how they react to them can influence how they will respond in class settings relative to non-moral cultural values. As you work with students, recognize the distinction between non-moral and moral cultural norms related to issues you use, and recall that differing premises result in differing conclusions, or in the same conclusion but for differing reasons.

Politics

Many teachers feel that topics such as democracy, socialism, communism, and totalitarianism should never be discussed in the classroom. Political ideologies can and do influence people's values, often in paradoxical and inconsistent relation to other values they hold. The political inclinations of your community may influence controversial issues in your classroom discussions. Popular political figures who have taken a position on a non-moral or moral issue may influence the values of the rest of the population. Other issues often are associated with various political ideologies and this "guilt by association" can result in confused discussion on the part of your students who are investigating the problem. It is necessary for you to understand the political influences operating in your community as you approach topics which may have been aligned with other ideological views. As before, being able to help your students distinguish between the components which are inherent in an issue and extraneous "baggage" which is carried along in the debate surrounding the issue is important for successful instruction and learning.

Geography and Ethnicity

America is not the "great melting pot" people once thought it to be. Ethnic origins and cultural differences have interacted with the geography of the United States to produce a highly diversified nation. The immigrant groups who settled the country collected in certain geographic localities throughout the country, bringing their cultural values and ethics with them. More than two hundred years later, or less, many of the ethnic imports have remained to provide a rich base of values. The area in which you teach will have its own unique ethnic flavor and values which will be part of your students' values repertoire. The interaction of the ethnic component with the geographic conditions of the country has resulted in a unique blend of views and life-styles which you must discern in order to understand your students, and to facilitate their education in the areas of values and controversial issues.

Religion

Along with other components of ethnic heritage, people brought their religious belief systems with them as they came to America. Religious beliefs are extremely powerful forces in the development of your students' values. A lack of awareness of the religious traditions and beliefs of the students in your classroom can be dangerous to you as you engage in discussions of controversial topics. Even non-controversial areas can be fraught with potential problems for the science teacher when religious values and beliefs are not recognized or understood by him or her. In the following chapter, the difficulties in resolving science instruction and religious beliefs will be examined in greater detail.

REFERENCES

[1]Taylor, Paul, Principles of Ethics: An Introduction, Dickenson Publishing Co., Encino, California, 1975.

[2]Ibid.

[3]Frankena, William, K., Ethics, 2nd Ed., Prentice-Hall, Inc., Englewood Cliffs, New Jersey, 1973.

7

SCIENCE VS. RELIGION: THE GREAT DEBATE

Introduction

Without making a single judgmental statement, a science teacher can suddenly become the focus of serious controversy. In this country, there are many regions in which mention of organic evolution or the age of the earth by a teacher can result in a variety of reprisals and admonishments. These can range from a call from a concerned parent to court proceedings and action for the teacher's dismissal. Regardless of what may be said about academic freedom, freedom of speech, and separation of church and state, the community pressure and reaction against certain scientific ideas in many localities in this nation remains strong and undiminishing.

It is not our purpose to encourage any of you to become martyrs in the cause of academic freedom, nor is it our intent to provide you with mechanisms for "sneaking" these topics into your curriculum. We think the situation where these strong local feelings prevail is quite serious, and that the short term efforts of any teacher or local group of teachers to "fight" these views would be counter-productive. Our purpose in this chapter is to help you understand the origins, facts, and fictions surrounding the issue, help you to recognize your rights and responsibilities as a science teacher in this area, and finally, to provide you with some suggestions for dealing constructively with the issue in your classes. We feel that you can do an effective job of dealing with certain problems without causing unnecessary controversy, but that to do this, you need to know about the issue, about your legal rights and professional responsibilities, and the community in which you are teaching.

Origins of the Conflict:

The antagonisms between science and the church seem to have begun about the beginning of the 16th Century, along with the development of science itself. The classic case of Galileo's difficulties with the church of his day resulted primarily from the belief of church leaders that Galileo's alternative explanation of the solar system challenged the authority of the church which had come to serve as the repository for much of all knowledge during Medieval times.[1] The development of the Renaissance, of which the sciences were an important component, was also viewed by the church as a turning away from God and the things of the spirit, and a concentration on the physical world.

This particular controversy eventually became reconciled and the next serious science-religious controversy occurred with the publication of Charles Darwin's book, *The Origin of Species*. Here the controversy centered on the questions which Darwin's book and ideas raised concerning the old classical interpretation of the Bible. The notion that man may not have been created along with the rest of the universe in the manner described in Genesis brought about a strong response from many people. This controversy extended from 1859 well into the 20th Century, and in effect, still persists today. The development of science caused great turmoil in Christianity. Many Christians have come to reconcile their religious views with the findings of science, but this is not true of all Christians. As a result of these influences, Protestant Christianity has been divided into two groups, which have commonly been called "modernists" and "fundamentalists." These groups cut across denominational lines and today there appears to be more distinctive differences between these groups within a denomination than between the denominations themselves.[2] Fundamentalism was an American movement which developed around the turn of the century, and among the issues of its original concerns was the theory of evolution.[3] The influence of fundamentalists was great and many states passed laws prohibiting the teaching of evolution in the public schools.[4] All of these laws have since been ruled unconstitutional, the last ones in Arkansas and Mississippi being stricken from state statutes in 1968. While prohibitory laws have been removed, fundamentalist activity in this area is not in decline and still proves to be a source of difficulty for science teachers.

Realities and Myths of the Conflict:

Realities: The public is concerned about science today. There are two areas of controversy which can be identified; controversy over the influence of science on society, and controversy over science as a way of knowing. Both of these areas are involved in the science-religion issue. In 1973, less than six years after repealing its anti-evolution statute, the state of Tennessee passed a new law requiring that any textbook which presented ideas concerning evolution must also give equal emphasis and space to alternative theories of creation, including the account found in Genesis.[5] The most recent tack taken by fundamentalist groups has been the argument of "equal time," and today, a relatively small number of dedicated people in this group have become highly organized to bring about such legislative actions.

Dorothy Nelkin, in her excellent account of the textbook controversy,[6] has identified a number of organizations which have been developed to promote the view of "special creation" which has been proposed as an alternative theory to evolution. The membership and leadership of many of these organizations have scientific credentials and have considerable financial support for their efforts.

To what extent does a conflict over science and religion seem to be present in the minds of students? The authors have asked this question and surveyed over 500 college and high school students in two states. Our findings indicate that between 52% and 77% of the groups surveyed had not resolved their personal conflicts between science and religion.[7] Our conclusion is that a significant number of students in the United States today do have unresolved conflicts in this area, and we believe this conflict does damage to both their interest in science and in religion.

Myths: As in the case of many controversial issues, many misconceptions and myths have developed which tend to obscure and complicate the issue. Among the primary myths in the science-religion controversy is that the scientific community and theological community are engaged in a pitched battle over it. As Martin Marty has pointed out, there is more conflict between theologians and other theologians than there is between the two groups.[8] For most theologians, this debate ceased to be relevant more than 70 years ago, and the vast majority of scientists are either apathetic to the issue or have reconciled the differences.

Another myth is involved with the problem of science as a way of knowing. The view that the Bible is relevant to scientific knowledge and that the theory of evolution is somehow a proposed description of reality indicates two problems; a lack of distinction between scientific and revelationary ways of knowing, and a misunderstanding of contemporary philosophy of science. Today, philosophers of science are unconcerned with the "truth" of scientific theories in the sense that they see such theories of science as models. The relevant question posed by these people (and most scientists) is: "Does nature in fact behave in a manner which is consistent with the model?" If the model does not fit, alternatives are created. The important point is that the model is "created" by people, and hence is no more "true" or "false" than any other model or human invention.

A final example of mythology attached to the controversy is that there is no basis for a dialogue between science and theology, that neither has anything to say to the other. Ian G. Barbour, in his book, *Issues in Science and Religion,* makes an eloquent case for the claim that they do have a need for dialogue. While Barbour contends that science and religion are "complementary languages," each having its own limitations, he also submits there are significant parallels in their methods. They both seek an integrated world view, and through successful dialogue, they yield a new view of nature, which could be mutually beneficial.[9] The realization that many present and future biomedical developments may pose profound moral and ethical problems seems to stand in support of Barbour's thesis.

Teaching Science: Rights and Responsibilities

Rights: As a science teacher, you should be aware of the fact that you have certain rights with respect to what and how you teach science. A variety of court decisions such as Epperson v. Arkansas, Daniel v. Waters, and Hendren v. Campbell make quite clear the unconstitutionality of both prohibitory and "equal time" legislation on this issue. In addition to these court decisions, many scientific and humanistic professional organizations such as the American Association for the Advancement of Science (AAAS), the National Science Teachers Association (NSTA), the National Association of Biology Teachers (NABT), and the American Humanist Association (AHA) have passed resolutions and issued statements reaffirming the rights of science teachers to teach evolutionary theory. Analysis of these court decisions, resolutions, and statements will provide you with a good understanding of your rights in this area. Also, several professional groups have defense funds available to member teachers who become involved in litigation on these issues.

In addition to your rights, the Biological Science Curriculum Study (BSCS) contends that students have "learner rights." They identify the following:

1. They have the right to study those controversial issues that are appropriate to their level of maturity. In general, those issues that learners of a particular age are beginning to form opinions about are appropriate to that maturity level.

2. They have the right to obtain relevant information freely.

3. They have the right to seek information in an atmosphere as free from bias, prejudice, and coercion as possible; bias, where it appears, should be clearly labeled.

4. They have the right to form and express opinions on controversial issues without jeopardizing their relations with teachers or peers.[11]

As we proceed to the next section on responsibilities, bear in mind the notion of "learner rights" for these rights have implications for the responsibilities of science teachers.

Responsibilities: Although we often fail to regard them as being as important as meeting our classes or filling out attendance forms, we, as science teachers, do have professional responsibilities with regard to what and how we teach science. First, we have a responsibility to be scientifically honest. That is, we have a duty to teach science, not as we or someone else might like it to be, but as it is, and in a manner consistent with the process and methodology of science. We have a responsibility for helping our students understand, to the extent they are able, that science is a way of knowing, not the way of knowing. In that same context, we should help students distinguish between scientific and other ways of knowing. Science defines our limits while religious ways of knowing seek to understand our purpose.

As science teachers operating in a democratic society, we also have a responsibility for conducting our classes in a manner which is consistent with the model valued and used by our society for addressing such issues. In this country, it is generally recognized that Americans handle their controversy according to accepted rules, and that this implies that there be a free examination of ideas and open inquiry into the issues themselves. If we are truly commissioned as part of our duties in the public schools to develop effective citizenship, then we should honor that commission by teaching in a manner consistent with effective citizenship, and its development.

Finally, we have a responsibility to deal honestly with student concerns related to science. In an age when we are sensitive to the need for addressing real problems in our classes, it would seem inconsistent to purposefully avoid this type of issue.

Suggestions to Teachers:

We are all responsible for using professional judgment in our teaching and many people will find a variety of ways to meet their professional responsibilities as science teachers. It is essential that we recognize our "rights" to teach something as being uniquely different from our judgment about how to proceed. We believe that each science teacher must carefully study both the issues surrounding science and religion and the context in which he or she is teaching before beginning to prepare classroom work on this area. Preparation and professional "wisdom" are important criteria to meet before you begin.

First, we suggest you carefully analyze your own teaching style or styles. If you tend to play the role of the authority, if you have a tendency to be dogmatic, or if your class rarely has the opportunity to engage in an open discussion of issues, you may be wise to begin by working on a more flexible style before beginning to work with controversy such as science and religion. Second, we strongly urge you to conduct a "self-analysis" such as the one described in Chapter 3 of this book before you begin. Third, we advocate your development of an awareness of the sociology of your community and its values. There are communities in this country in which your best efforts in this area will be totally counter-productive simply because community norms and religious values will not tolerate an open discussion. If you find you are teaching in such a community, you may be wise to stick to other issues and develop class work which may be less emotionally loaded. Fourth, identify community resources which may provide support for such class work. Again, this suggestion requires good judgment on your part and demands that you have a good understanding of the community in which you teach. Local resources may be people in the community to whom you can refer students, should you feel it unwise to bring up such issues in the classroom. These resources may be useful in helping you learn about your community, and may be willing and able to meet with your classes to provide other information for them.

As you develop class activities in this area, you may wish to consider the following suggestions: First, set guidelines (preferably with your students) as to how such classroom discussion should occur. At such a planning stage, you should be sensitive to the "learner rights" outlined. You may also recognize that your students are too emotionally involved with the issue to pursue it any further! Second, identify your resources. As it is possible, make use of community resources and carefully examine commercially prepared resources which usually deal with the issue in a balanced manner. Third, keep communication lines open. It is critical that good communications between the schools and the community be operating, for should your efforts appear to be clandestine, you may have difficulties. Fourth, select teaching strategies such as clarification and analysis which are described in Chapter 8 to develop classroom interaction and discussion of the issues. These two strategies are particularly suitable when issues have the potential for creating emotional reactions in your students. Fifth, be sure to recognize your own biases about this issue and always distinguish for your students which statements you make are your opinions and biases, and which are statements of facts or findings of others.

Finally, respect the limitations of your students. Many students may find it impossible to approach these issues without emotional reactions taking over. Under such conditions, these students should be allowed the opportunity of not participating in the class activity. This is a sensitive area, but an important one for science teachers. To deal in a professionally effective manner with it is probably one of the greatest challenges you can accept as a science teacher. Do not mistake timidity for wisdom, but also recognize your professional responsibilities.

REFERENCES

[1]Butterfield, Herbert, The Origins of Modern Science, Rev. Ed., The Free Press, New York, New York, 1965, p. 68.

[2]Braden, Charles S., The World's Religions, Revised, Abingdon Press, New York, New York, 1954, p. 212.

[3]Nelkin, Dorothy, Science Textbook Controversies and the Politics of Equal Time, The MIT Press, Cambridge, Massachusetts, 1977, p. 13.

[4]Ibid., p. 14.

[5]LeClerq, Frederic S., The Constitution and Creationism, American Biology Teacher, March, 1974, Vol. 36, No. 3.

[6]Nelkin, Dorothy, Science Textbook Controversies and the Politics of Equal Time, The MIT Press, Cambridge, Massachusetts, 1977, Chapter 5.

[7]Rusch, John, and Barman, Charles, "Helping Students Distinguish Between Scientific and Religious Ways of Knowing," paper presented at the NSTA Annual Meeting, Philadelphia, Pennsylvania, March, 1976.

[8]Marty, Martin E., "Science Versus Religion: An Old Squabble Simmers Down," Saturday Review, December 10, 1977, pp. 29-35.

[9]Barbour, Ian G., Issues in Science and Religion, Prentice-Hall, Inc., 1966.

[10]National Association of Biology Teachers, A Compendium of Information on the Theory of Evolution and the Evolution Controversy, 1978.

[11]BSCS Committee, "Issues and Controversy in Teaching Science," The Biological Sciences Curriculum Study Journal, Vol. 1, February, 1978, pp. 33-37.

8

DEVELOPING VALUES IN YOUR SCIENCE CLASSROOM

Values are being continuously taught and developed in classrooms all over the United States at this time. Most values being promoted are those held by teachers and administrators, and students are either incorporating these values into their own values patterns, or they are reacting against these values. The values are almost invariably being taught didactically by inculcation, or they are presented more subtly, through the behavior of the teacher.[1] All of this may be either good or bad, depending upon the people involved, their values, methods and behaviors. Nevertheless, we propose that there is reason to believe that using planned strategies to help students develop their own values is to be preferred over unplanned, poorly devised, and spontaneous classroom sessions which often occur in an emotional or pedantic context.

According to Sidney B. Simon, students who are exposed to clarification lessons are more able to critically analyze problems and also demonstrate the ability to follow through on their decisions.[2] We believe you can do an effective job of helping students develop their values in your science classroom and that this can positively influence their lives. We also believe that you can do this in a humane and democratic context, recognizing and respecting the rights and existing values of everyone in that classroom. We are convinced that when you do this in the context of contemporary human problems related to the science being studied in your classes, you gain the added advantage of demonstrating the relevance and importance of scientific literacy and involved citizenship to your students. We further believe that this will help develop more motivation and interest in science among your students because they can incorporate and relate that science more completely into their own lives.

In this chapter we will attempt to identify some areas from which you can draw your own values lessons in science, we will provide you with some examples of how to develop certain types of values lessons, and we'll suggest some ways of incorporating the activities in the second part of this book into your classes.

Resource areas for value-laden issues in science:

Generally, there are three basic areas from which controversial science issues can be derived: controversies arising within science disciplines themselves; controversies arising from concerns over the effect of science and technology on people's lives; and finally, controversies arising over science as a way of knowing. Each of these areas have potentially useful materials which science teachers can put to effective use in their classrooms.

There are numerous examples of controversies within science disciplines which can be effectively developed. Many science teachers have used these issues successfully for years as means of helping their students conceptualize the humanness of scientists and as a way of gaining insight into how the disciplines themselves have developed. In the area of Biology, there are many classic controversies, such as the Darwin-Lamarck debate, the controversy over Pasteur's germ theory, and more recently, the debate over the development and control of recombinant DNA research. The Earth Sciences have also experienced their share of controversy throughout their history. The old arguments between the "Neptunists" and "Plutonists," the "Catastrophists" and the "Uniformitarians," can be very nicely adapted to classroom use. Other problem areas in Geology, such as the development of Global Tetonic Theory in the past few decades can easily be researched by examining *Geotimes* articles of the late 1960's and early 1970's.

In the physical sciences, the arguments over "phlogiston," the "ether," the wave or particle nature of light, and the concerns of World War II scientists, as evidenced by periodicals such as the Bulletin of the Atomic Scientists, are but a few examples of human concerns and sometimes bitter conflicts within the scientific community. These can serve to help you give students a more human view of the nature of the scientific enterprise. Although such issues have often been presented in a lecture format, or by way of a case history or case analysis strategy, we suggest that other strategies such as role-playing and analysis can be quite effective, especially when you are attempting to get greater class involvement, and wish to develop the valuing dimensions of the issues.

In Chapters 5 and 6 of this book we have cited a number of possible issue areas which arise out of controversies over the effect that scientific and technological change have had on people's lives. It is difficult to pick up a newspaper, popular magazine, or science periodical today which does not provide some type of information on these issues. Among the most easily accessible resources for science teachers are the following professional journals, all of which offer either a special section on such issues or which maintain a relatively constant flow of articles relating to such issues.

The Science Teacher

The American Biology Teacher

The Journal of Chemical Education

The Physics Teacher

The Journal of Geological Education

Science

A careful perusal of the issues of the past 5 years of these journals will supply you with a wealth of ideas and topics which can be continuously supplemented by current periodicals, newspapers, television, and radio.

The area of controversy over science as a way of knowing has been discussed to some extent in Chapter 7, but primarily from the context of the science-religion controversy. It can also be approached, and perhaps more easily, from the perspective of the history and philosophy of science. While this area can be quite abstract and much of it may lie beyond the abilities, background, or developmental level of your students, it is possible to find many ways of developing a meaningful understanding of the nature of science. Acquiring the concepts that science is a way of constructing reality and not a way of describing it; that the "laws of nature" are indeed made by man; and that scientific theories are embraced or rejected by the scientific community more on the bases of their potential for allowing new avenues of investigation, and, perhaps their esthetic value, than on the basis of their "truth" are important objectives for any secondary science curriculum. Moreover, these concepts can set the stage for a much more rational approach to more controversial issues such as the science-religion controversy.

In this area, you will have to go to a wide variety of books which deal with the history and philosophy of science. Also, biographies of great scientists often provide a wealth of information which can be incorporated into class activities and discussions. Some of these books are cited in the bibliography at the end of this book.

Developing Values Lessons for Your Classroom:

You now have some idea of general areas and references from which to draw materials for the development of your own values lessons. There are several steps which we would suggest you follow in the actual planning of a values lesson.

1. Analyze each unit or chapter that you plan to teach for ideas which might be appropriate for developing a values lesson which fits the materials in the chapter. For example, an earth science chapter on rivers and streams would provide the context for values lessons on human choices related to building homes on flood plains, and the channelization of streams and its consequences. A biology unit on Botany may provide a good basis for discussing food production, the "green revolution" and world hunger.

2. Get input from your students regarding areas of social concern which they might like to investigate in class. Often, students have very good suggestions for discussion topics, and may even be able to help you in the planning and resource gathering for the lesson.

3. Select the appropriate values education approach for each individual lesson. This, of course, implies that you have some well-developed objectives for the unit or chapter which deal with values development. If the main purpose of your lesson is to clarify existing student values, if it is to initiate student action on a problem, or if it is to analyze an issue, you will need to make decisions to use strategies which are appropriate to your objectives.

4. During the lesson itself, keep some notes of issues that are raised along with the main issue, make careful observations of student reactions, and generally evaluate the effectiveness of the lesson. Keeping track of lessons in this manner can allow you to develop spin-off lessons, and to make modifications in your original lesson more easily.

To help you develop a better concept of some various strategies for values lessons, we will supply you with four examples of different kinds of values lessons. These lessons are not necessarily suggested for use in your classroom, but should provide you with an idea of how certain types of objectives can be reached with certain strategies.

I. Moral Development - A solution to a moral dilemma may be solved through personal feelings and conscience.

Example: "What Would You Do?"

A genetic counselor's services have been sought by a couple who have a child with a serious genetic defect. The parents want to know their chances of having another child with the same defect. After some examination, the counselor finds that the genetic defect is not present in the husband's genes. Since both mother and father must carry the genetic information which will cause the defect in their offspring, the counselor knows that the child with the genetic defect must have been fathered by someone other than the husband.

Should the counselor divulge this information to the husband?

II. Clarification - The main purpose is to assist students in developing the "valuing process."

Example: "Electricity and You"

List ten electrical appliances that you or your family use at home.

1.	6.
2.	7.
3.	8.
4.	9.
5.	10.

Circle those appliances that you feel you or your family could do without.

1. How many appliances on your list are essential for you and your family?

III. Action Learning - This approach provides opportunities for your students to act on their values.

Example: The "Electricity and You" activity could be expanded in the following manner.

2. Make an attempt for one week to live without the electrical appliances you circled.

3. Keep a record of your progress for each day.

IV. Analysis - This method provides opportunities for students to critically analyze certain social values issues.

Example:

Question: Should human medical experimentation, such as the example cited below, be allowed?

Purpose of Experiment: To determine if the new antibiotic B is as effective in treating Staph infection as antibiotic A. (Fact: Antibiotic A has been proven effective in treating Staph infection for 20 years.)

Experimental Design:

1. Randomly divide 100 patients suffering from the same Staph infection into two groups of 50.
2. Group I -Treat this group with antibiotic A.
   Group II - Treat this group with antibiotic B.
3. Subjects in this experiment will not be informed that they are participating in an experiment.
4. All subjects will think that they are being treated with antibiotic A.

In groups of four or five, compile a list of reasons why you feel experiments of this type should or should not be allowed. Time will be allotted for each group to discuss their list with the rest of the class.

Integrating Values and Controversies into Your Curriculum:

Earlier in this chapter we alluded to the fact that an analysis of your existing course objectives was an important step in using values lessons in your science classrooms. Actually, it is essential. We do not advocate teachers' using "values activities" unless the activities are consistent and defensible in terms of a carefully worked out science curriculum. We feel that any science curriculum should have, in addition to science content and science process goals, goals related to the cultural implications of science and the nature of the scientific enterprise. It is in these last two areas that we feel values lessons may be most effective. If your existing curriculum does little with these two major goal areas, it is possible that you will need to re-think your curriculum before planning values lessons.

This suggestion results in another commonly asked question: "How will I get this material into my course when I can't complete my present syllabus?" We agree that some things will probably have to be omitted from your present curriculum to accommodate the addition of values lessons. This, however, does not necessarily mean that the quality of your course with respect to content and process objectives must suffer. We believe that the addition of values-related objectives to your science curriculum will enhance the course, and provide students with more motivation for developing those concepts than they previously had. Whenever you carefully scrutinize an existing science course, you can usually identify areas and topics which have only marginal significance in terms of the main goals of the course. Professional judgment is important in deciding how and what to delete and add, and that particular responsibility is always in the hands of the teacher. Remember, there are many ways to achieve the goals of a given course, some of which are more efficient than others, and by comparing your teaching style, the nature and abilities of your students, and your resources to your existing goals, you may find many ways to streamline your course to accommodate values lessons without doing violence to your content and process goals.

The next question raised in most discussions dealing with curriculum revision is related to evaluation. As we discussed earlier in Chapter 5, it is quite possible to identify objectives in the affective domain, and, by making careful observations of your students' behavior in the sessions and following the sessions, you should be able to obtain evidence for determining whether or not your objectives have been achieved. Again, you will have to use professional judgment in evaluating your objectives in these areas, and the means you use for doing so are only limited by your imagination and creativity.

Finally, we'd like to address the question of the use of controversies in terms of whether or not they are necessary to develop values objectives. First, we don't believe it is absolutely necessary to use controversial issues in your classroom to achieve objectives in science. There are a number of alternatives to the use of controversial issues which can be employed to achieve a particular objective. If you feel uncomfortable with the idea of using a controversial issue in your classroom, you probably shouldn't; at least, not until you feel you have developed sufficient skill in working with less controversial issues related to values education.

We do think that controversial issues are part of real life however, and therefore, they are issues around which your students can more effectively interact. If we wish to make our science classes as relevant as possible, we need to seriously consider using such issues. Secondly, although they are not necessary, they are probably the only way in which you can pattern your classes after real life situations, the type of situations in which your students will find themselves after leaving secondary school. To the extent that you feel such experiences are a necessary part of the science education of your students, you are responsible for dealing with those issues. The choices, values, and judgments you make about your own teaching are yours. You must decide.

REFERENCES

[1]Herron, J. Dudley, "Implicit Curriculum . . . Where Values Are Really Taught," *The Science Teacher,* 44(3): 30-31, March 1977.

[2]Simon, Sidney B., et al., Values Clarification, Holt Publishing Company, New York, 1972.

9

ACTIVITIES

The following activities represent about five years of development and field testing with students and teachers throughout the United States and Canada. The feedback from each of these individuals has been instrumental in developing new lessons and revising old ones. To all of the people who provided input into the efficacy of these lessons, we would like to offer a special thanks.

We realize that some of the content in these activities may be somewhat obsolete by the time you review them. Therefore, if necessary, we encourage you to modify these lessons to accommodate your individual needs.

In addition, these are only sample activities. They are not presented as a developed sequential curriculum. The best activities and lessons are those which you develop yourself for your classes.

Human Medical Experimentation

Note to the teacher: We suggest that you provide time for your students to discuss their answers to the questions in this activity. A possible follow-up lesson to this exercise could be "Medical Investigation and Child Subjects."

The following study was cited in *The New England Journal of Medicine:*

> "Live cancer cells were injected into 22 human subjects as part of a study of immunity to cancer. According to a recent review, the subjects (hospitalized patients) were merely told they would be receiving 'some cells'—the word 'cancer' was entirely omitted . . ."[1]

The investigation described above is one of one hundred or more that have been conducted on human subjects in the United States and England. In all of these experiments the participants were either misinformed or not informed at all of the conditions and/or hazards of the investigations.[2]

This type of scientific endeavor raises many moral and ethical questions. Please read the following questions and explain your feelings toward them.

1. To what extent do you feel a person taking part in a medical experiment should be informed of the circumstances and conditions of the experiment?

2. What makes a medical experiment ethical—the procedure that is used or the results it yields?

3. Would you take part in a medical experiment if it could be beneficial to mankind? Explain your answer.

4. To what extent should parents be allowed to decide whether or not their children can participate in a medical experiment? Explain the rationale for your answer.

[1]Beecher, Henry K., "Ethics and Clinical Research," *The New England Journal of Medicine,* 274:1354-60, 1966.

[2]Ibid.

Medical Investigation and Child Subjects

Note to the teacher: You may want to direct Parts II and III separately. We suggest that you provide an opportunity to discuss each part of this lesson.

Part I

The following hypothetical situation was part of a study conducted in the United States.[1] This and other similar situations were distributed to a random sample of adults to determine their feelings towards specific types of medical experimentation.

Please read the following paragraph:

A researcher plans to study bone metabolism in children suffering from a serious bone disease. He intends to determine the degree of appropriation of calcium into the bone by using radioactive calcium. In order to make an adequate comparison, he intends to use some healthy children as controls, and he plans to obtain the consent of the parents of both groups of children after explaining to them the nature and purposes of the investigation and the short- and long-term risks to their children. Evidence from animals and earlier studies in humans indicated that the size of the radioactive dose to be administered here would only very slightly (say, by 5-10 chances in a million) increase the probability of the subjects involved contracting leukemia or experiencing other problems in the long run. While there are no definitive data as yet on the incidence of leukemia in children, a number of doctors and statistical sources indicate that the rate is about 250/million in persons under 18 years of age. Assume for the purpose of this question that the incidence of the bone disease being discussed is about the same as that for leukemia in children under 18 years of age. The investigation, if successful, would add greatly to medical knowledge regarding this particular bone disease, but the administration of the radioactive calcium would not be of immediate therapeutic benefit for either group of children. The results of the investigation may, however, eventually benefit the group of children suffering from the bone disease. Please assume for the purposes of this question that there is no other method that would produce the data the researcher desires. The researcher is known to be highly competent in this area.

Part II

Now assume the role of a parent whose request is being sought to allow his/her child to participate in the experiment outlined above. Check only one of the items that you would consider acceptable for your approval of the proposed investigation.

( ) 1. If the chances are 1 in 10 that the investigation will lead to an important medical discovery.

( ) 2. If the chances are 3 in 10 that the investigation will lead to an important medical discovery.

( ) 3. If the chances are 5 in 10 that the investigation will lead to an important medical discovery.

( ) 4. If the chances are 7 in 10 that the investigation will lead to an important medical discovery.

( ) 5. If the chances are 9 in 10 that the investigation will lead to an important medical discovery.

( ) 6. Place a check here if you feel that, as the proposal stands, the researcher should not attempt the investigation, no matter what the probability that an important medical discovery will result. (IF YOU CHECKED HERE, please explain):

What is the rationale for your above answer?

Part III

Next, assume the role of a doctor who is in charge of granting approval of all of the medical experimentation that is conducted at a specific hospital. What level of chance would you consider acceptable to approve this experiment? Check only one item.

( ) 1. If the chances are 1 in 10 that the investigation will lead to an important medical discovery.

( ) 2. If the chances are 3 in 10 that the investigation will lead to an important medical discovery.

( ) 3. If the chances are 5 in 10 that the investigation will lead to an important medical discovery.

( ) 4. If the chances are 7 in 10 that the investigation will lead to an important medical discovery.

( ) 5. If the chances are 9 in 10 that the investigation will lead to an important medical discovery.

( ) 6. Place a check here if you feel that, as the proposal stands, the researcher should not attempt the investigation, no matter what the probability that an important medical discovery will result. (IF YOU CHECKED HERE, please explain):

1. Compare this response to the one in Part II. Did it remain the same or change?

2. What is the rationale for your answer to Part III?

REFERENCES

[1]Barber, Bernard, "The Ethics of Experimentation with Human Subjects," Scientific American, 234(2): 25-31, Feb. 1976. Reprinted with the author's permission.

Society, Science, and Technology

Note to the teacher: This activity is designed to initiate discussion about the ways in which science, technology, and society are interwoven. We suggest you provide time for the students to discuss, in small groups or as a class, their answers to this activity.

Directions for 1 to 3: Use a check mark to indicate your feeling relative to the extremes listed.

1. The term "society" means to me:

a. optimistic	___ : ___ : ___ : ___ : ___ : ___ :	pessimistic
b. positive	___ : ___ : ___ : ___ : ___ : ___ :	negative
c. complete	___ : ___ : ___ : ___ : ___ : ___ :	incomplete
d. severe	___ : ___ : ___ : ___ : ___ : ___ :	lenient
e. strong	___ : ___ : ___ : ___ : ___ : ___ :	weak
f. tenacious	___ : ___ : ___ : ___ : ___ : ___ :	yielding
g. active	___ : ___ : ___ : ___ : ___ : ___ :	passive
h. excitable	___ : ___ : ___ : ___ : ___ : ___ :	calm

2. The term "science" means to me:

i. intentional	___ : ___ : ___ : ___ : ___ : ___ :	unintentional
j. stable	___ : ___ : ___ : ___ : ___ : ___ :	unstable
k. rational	___ : ___ : ___ : ___ : ___ : ___ :	intuitive
l. cautious	___ : ___ : ___ : ___ : ___ : ___ :	rash
m. valuable	___ : ___ : ___ : ___ : ___ : ___ :	worthless
n. mobile	___ : ___ : ___ : ___ : ___ : ___ :	stationary
o. humorous	___ : ___ : ___ : ___ : ___ : ___ :	serious
p. kind	___ : ___ : ___ : ___ : ___ : ___ :	unkind

3. The term "technology" means to me:

q. unscientific	___ : ___ : ___ : ___ : ___ : ___ :	scientific
r. intelligent	___ : ___ : ___ : ___ : ___ : ___ :	dumb
s. pretty	___ : ___ : ___ : ___ : ___ : ___ :	ugly
t. frightening	___ : ___ : ___ : ___ : ___ : ___ :	secure
u. round	___ : ___ : ___ : ___ : ___ : ___ :	square
v. real	___ : ___ : ___ : ___ : ___ : ___ :	make-believe
w. varied	___ : ___ : ___ : ___ : ___ : ___ :	monotonous
x. impersonal	___ : ___ : ___ : ___ : ___ : ___ :	personal
y. good	___ : ___ : ___ : ___ : ___ : ___ :	bad
z. unnecessary	___ : ___ : ___ : ___ : ___ : ___ :	necessary

4. List two questions concerning science-linked social problems you think are important.

a.

b.

5. Please indicate the extent of your agreement or disagreement with each of the following statements by circling the appropriate number at the right of each statement.

(1) SA - I strongly agree with the statement.
(2) A - I agree with the statement.
(3) U - I am undecided about the statement.
(4) D - I disagree with the statement.
(5) SD - I strongly disagree with the statement.

	SA	A	U	D	SD
a. Man's relationships to things and people are increasingly permanent.	1	2	3	4	5
b. Art, today, is instant computer created, impermanent.	1	2	3	4	5
c. What is built must be easily dismantled.	1	2	3	4	5
d. Genetic supermarkets are desirable.	1	2	3	4	5
e. Individuals are becoming alienated and anxiety ridden.	1	2	3	4	5
f. We may soon be able to duplicate humans genetically, down to the last detail.	1	2	3	4	5
g. We should control the code which determines skin color.	1	2	3	4	5
h. We cannot expect two people to grow together, change together, and complement each other till death do them part.	1	2	3	4	5
i. The control of intellectual functioning by drugs and electrical stimulation is desirable.	1	2	3	4	5
j. Technology can be used to build a decent, democratic, and human society.	1	2	3	4	5
k. Technology should be controlled more.	1	2	3	4	5

Automobile Safety Restraints

Note to the teacher: This activity is designed to be used after studying Newton's first law of motion. Allow the students time to discuss their answers in small groups or as a class. A suggested follow-up activity would be to have the class poll other students, faculty, or other community members on the following issues:

1. Review the relationship between Newton's first law and automobile seat belts. How are head restraints in cars related to Newton's first law and how do they protect people?

2. Do you have seat belts in the car in which you drive most? Do you use them? Why or why not? What are the possible consequences of your choice to use or not use seat belts?

3. Some members of Congress would like to get a law enacted requiring the use of seat belts. What would you say to a Congressperson if you were asked for your feelings about such a law?

4. Insurance companies claim that many lives now lost in highway accidents could be saved by the use of passenger restraining devices. These devices could be active or passive. Seat belts which require buckling are an example of an active restraint. Air bags which inflate on collision are considered passive restraints. Air bags could increase the cost of an automobile by $500 or more. Which of these devices would you prefer in your automobile if you had to make a choice?

Energy Choices in a Crisis

Note to the teacher: This activity is designed to initiate discussion on energy usage and possible energy shortages. Allow enough time for students to respond to the items and for class discussion.

In 1973-1974, the Arab oil embargo caused long lines at gasoline stations, high fuel bills, and general disruption of normal day-to-day activities. Political situations in countries which export oil could once again cause fuel shortages around the world.

Please respond to the following items concerning energy.

1. Write a short paragraph describing how you perceive your role in an energy crisis.

2. Which of the following would you prefer?

a. 55 miles per hour	or	the "old" speed limits
b. driving alone	or	car pooling
c. mandatory fuel rationing	or	voluntary conservation
d. an electric blow-dryer	or	wet hair
e. Cadillac El Dorado	or	Chevy Chevette
f. thermostats at 65° in winter	or	72° in winter
g. taking a trip	or	vacationing at home

3. How do preferences above parallel or oppose the first paragraph you wrote concerning your perceived role in an energy crisis?

4. How do you feel about manufacturers continuing to develop small appliances which consume electricity?

5. What sacrifices would you be willing to make in order to conserve energy if you had to?

6. List two energy-using devices your family could least do without.

7. List two energy-using devices your family could easily live without.

The Metric System

Note to the teacher: Almost all science classes have a unit on measuring. Much controversy surrounds possible plans for a United States conversion to the metric system. A suggested follow-up activity is to have students poll local industrial and business leaders about how converting to the metric system would affect their industries and businesses.

The United States is now the only major country in the world which is not officially using the metric system of measurement. Laws have been introduced and defeated in Congress to have the United States convert to the metric systems. Opponents say it is too costly and unnecessary.

1. If you were polled about your feelings on converting to the metric system, how would you respond?

2. Make a listing of some advantages and disadvantages of the United States converting to the metric system.

3. Some critics of conversion to the metric system say that other countries should convert to our system of measurement because the U. S. is the largest producer and user of the world's goods. What are your personal feelings about this criticism?

4. Why do you think some people are so reluctant to change to the metric system?

5. How receptive are you to new ideas?

6. How receptive or resistant to change are you?

Nuclear Power Stations

Note to the teacher: This activity is designed to be used after studying units on energy or nuclear physics. Allow time for students to discuss their feelings about the issues. Follow-up activities might include polling community members about their feelings on nuclear generating plants.

The small town of Palo, Iowa, has been in the news since the late 1960's. In Palo, the Duane Arnold Energy Center, a nuclear power station, has been the site of controversy, numerous protests, and threats. The plant was closed for a while to repair cracks in water pipes. During that time, company officials received letters threatening sabotage, terrorism, and kidnapping if the plant were brought back into operation.

Strangely, most of the protesting and fear came from activist groups and not from the residents of the small town. The town's residents were largely upset about the constant publicity about the nuclear facility.

1. How would you feel about living close to a nuclear power generating plant?

2. Do you feel that nuclear power plants should continue to be developed? Why or why not?

3. What are the advantages of further expansion of nuclear generating stations? What significant social effects would be realized?

4. What are the disadvantages of further expansion of nuclear generating stations? What significant social effects would be realized?

5. Why would some activist groups react violently to nuclear power plants?

6. What methods would you use to support nuclear facility expansion?

7. What methods would you use to oppose nuclear facility expansion?

Town Council Meeting

Note to the teacher: This activity is an exercise in role playing. It is suggested that this activity be used after your class has studied about nuclear energy production. Try to select students for roles you feel would be opposite or different from the students' actual viewpoints.

You are at a special session of the Green Acres Town Council. This is the last of a series of special meetings being held in the Green Acres area to deal with the controversial issue of constructing a nuclear electric generating station on the edge of town. The construction site is along the beautiful Trout River.

Tonight the board is to come to a decision on a recommendation to be made to the Public Utilities Commission about the construction of the Nuclear Generating Station.

Council President Joe Smith has allotted time on the agenda for both sides to present final arguments to the Council. Time has been allotted for concerned townspeople to question the Council and the representatives.

AGENDA

15 - 20 Minutes: Presentation of final arguments to the Council.

15 - 20 Minutes: Discussion is open to the townspeople.

5 - 10 Minutes: Board votes and makes a recommendation.

****

President - Joe Smith: Mr. Smith is a strong individual and wants to run an orderly meeting to insure both sides a fair chance to present their viewpoints. Joe also feels that it is important for the townspeople to enter into the discussion.

Ms. Jane Adams: She is concerned with the effect of radiation on the community's children and the economic impact on the town.

Dr. John Marcellus: Dr. Marcellus served in postwar Japan with victims of the A-bomb blasts, but has not hinted as to how he stands on the current issue. He is well aware of the effects of nuclear disaster.

Mr. William Alton: Mr. Alton is a member of the Chamber of Commerce and feels the construction of the generating station will be a boost to the area's sagging economy.

Mr. Fred Fair: Mr. Fair is a level-headed individual who is willing to hear both sides. He thinks a compromise could be worked out in choosing an alternate site in the area, and using cooling towers to reduce discharge water temperature.

****

Scientists present:

Dr. Sam Newton: Corps of Army Engineers. He says this is the best possible site.

Dr. Rachel Wayne: Nuclear physicist working for Better Services Inc., the State Utility Company. She uses statistics that the chances of a nuclear disaster are one in a thousand.

Dr. Frank Field: Representing the Sequoia Club. He adamantly opposes nuclear power plants. He argues that the safety equipment has never really been tested, and cites that the A.E.C. considers a disaster one in which 50,000 or more must be killed or injured.

Mrs. Sally Nelson: A member of a local environmental group. She is also the local biology teacher.

****

Townspeople: Interested in the possible environmental problems, the economic uplift, damage by floods, the effect on human and livestock health, and other impacts of the nuclear generating station.

Role Playing Exercise: Should Eight Corners Include Creation in Science Instruction?

Note to the teacher: This role playing activity can be useful in initiating a discussion about the evolution-creation controversy. If possible, try to select students to assume roles that hold differing views from their own. It is also recommended that you allow ample time for your students to prepare for their roles and to review the common rules of order for public meetings.

You are at a special session of the Eight Corners School Board. The reason for the dilemma is, John Newt, the biology teacher at Eight Corners High School, is insisting on teaching evolution as the only means of explaining the origin of life. Some of the board members and townspeople feel that he should also present the Creation Theory.

The meeting begins with the school board members discussing their feelings about the issue. After the school board member have presented their views, they call Mr. Newt to explain his actions in class. When Mr. Newt is finished with his defense, any person from the town can come up and give his/her opinion.

The meeting is adjourned after everyone has said his piece and the school board has made a judgment on the matter.

SCHOOL BOARD MEMBERS: The philosophy of each school board member is presented below; however, each individual is open to compromise.

PRESIDENT: Jane Smith - She is a strong individual who feels both theories should be presented.

Rev. Genesis - He is not at all in favor of teaching evolution. He feels that Creation is the only way to explain our Earth's development.

Dr. Sally Newton - She has a Ph.D. in biochemistry and feels that religion should be left out of schools. She obviously favors evolution.

Joe Nelson - He has no opinions of his own. He is easily swayed by whomever is speaking.

Bill Adams - He is very much in favor of presenting Creation. He tends to favor omitting the teaching of evolution.

Mr. John Newt - He feels that evolution should be taught in school and the Bible should be taught in church.

TOWNSPEOPLE: Any person in class who does not have one of the above roles can express his/her opinion on this issue after the role players have made their comments.

Weather Modification and Human Conflicts

Note to the teacher:

This activity is intended to involve the students in thinking about the consequences of the application of technology and the legal needs which may result from such technologies.

Make sure you provide enough time for the groups to thoroughly develop responses to the three activities posed in this exercise. Some students may need some help in getting started making lists of questions, the answers to which would help them in defending or attacking Far West County. You may wish to provide some examples to get them started.

At the time of class discussion, put all the questions and compromises suggested by the various groups of students on the board before evaluating any of them. Have the class criticize the questions and compromises.

The last three questions of the activity should also be discussed by the entire class after each student has attempted to answer them. While there may be no "right" or "wrong" answers to these questions, they may provide a basis for further study of this kind of issue.

A three-county area in eastern Colorado is experiencing a severe drought. Since most of the people in these three counties are wheat famers or people who provide services to the wheat farmers, the economies of all three counties are in serious trouble. Far West County, lying to the west of the other two drought-stricken counties, holds a public meeting on the drought. At the meeting it is suggested that a commercial cloud-seeding company be hired to seed clouds over the county in an effort to make it rain. The county board appropriates $10,000 to hire this company and they begin seeding clouds. Within a week after they begin, clouds of the right type appear to the west of Far West County. They are moving east and will soon be overhead. The company prepares its planes to seed these clouds when they come over the county. They feel the chances of producing rain from these clouds is about 90%. They seed the clouds and an extremely successful rainfall of about 1.5 inches is produced. The following day, Far West County is sued by Middle and Far East Counties which lie directly east of Far West County. They are also suffering from the drought and claim in their suit that Far West County interfered with the natural processes which would have brought them the rain which they desperately need. In effect, they claim that Far West County "stole" their potential rain.

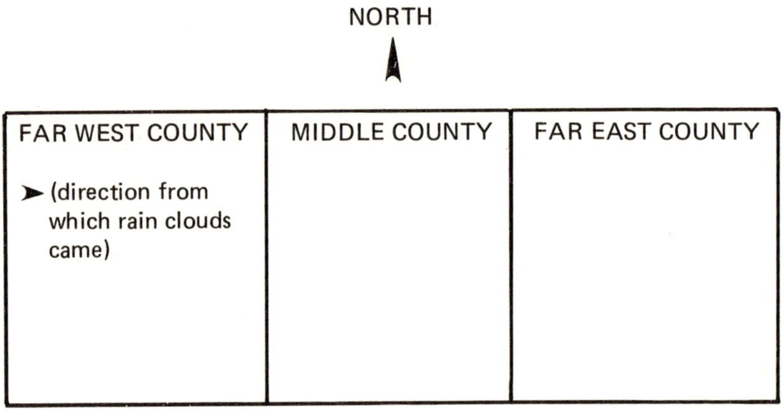

Working in groups, do the following activities:

1. Identify the kinds of information you would need to act as the District Attorney of Far West County. Your job is to defend Far West County and prove that it did not "steal" rain from Middle or Far East County. Make a list of questions which you can give to your assistants which could help you in the case. (In other words, what would you have to know to defend Far West?)

2. Identify the kinds of information you would require to serve as the District Attorneys of Middle and Far East Counties. Your job is to prove that Far West County did indeed steal rain from your counties and that they should pay damages. Make a list of questions which you can give to your assistants to help you in your case. (What would you have to know to win your lawsuit?)

3. As District Court Judge, you will have to rule on the lawsuit brought by Middle and Far East Counties against Far West County. If neither side makes a totally convincing case for or against the lawsuit, you will have to make some sort of compromise decision. What kinds of compromise decisions could you possibly make which could satisfy all three counties?

Discuss with the rest of the class the lists and compromises you developed for each of the three questions above.

Now answer the following questions:

1. Do you think that anyone should be allowed to hire a "rain-maker"? Why?

2. Should there be governmental regulations on how weather modification techniques can be applied? What kinds would you suggest, or why are you opposed to such regulations?

3. What kinds of questions should be researched by weather modification scientists to help avoid the kinds of problems which developed between these three counties?

People, Pollution, and Raindrops

Note to the teacher: This activity can be used as a basis for class discussion or as a role-playing exercise. Allow some time for students to read and react to the issues before starting class discussion.

Sudbury, Ontario, is the home of the richest nickel deposit in the world. A smelting operation is similarly located in Sudbury. Scientists and environmentalists are increasingly concerned about the smelting operation's contribution to the problem of acid rain. It is claimed that suplhur oxides from the smelter are combining with moisture in the air to form sulphuric acid. This acid lowers the pH of rainwater which in turn lowers the pH of lakes surrounding Sudbury. One scientist estimates that as many as 400 lakes within a fifty-mile radius of Sudbury have very few or no fish because of the acid rain.

If regulations were imposed, the nickel and smelting operations would be forced to close because it would be too expensive to clean up their processing procedures. Many people would lose their jobs if this were to happen.

If you had the authority to close the industry, what would you say to the townspeople? What values do you see at conflict here? What do you think the following individuals might say about the closings?

a. A supermarket manager whose livelihood depends on the nickel industry

b. An employee of the smelting plant who has a family to support

c. A local guide who is losing money because people no longer come to the area to fish

d. The vice-president of a company which uses nickel as a raw product in its industry

e. A leading political figure of the town who points out the benefits the nickel industry brings to the town

f. An independently employed local resident, not dependent on the industry

g. Other townspeople

Acid Rain

Note to the teacher: This activity can be used in conjunction with environmental studies, with chemistry classes, or general science classes studying acids and bases. Follow-up activities could include actual student testing of pond, river, and lake waters. Studying the behavior of various species of animal life under differing environmental acidity conditions could also be a spin-off activity.

Salamanders dying in the northeastern United States! Fish dying in northern Minnesota lakes! All because of rain? According to the U. S. Environmental Protection Agency, it's entirely possible. An increased rainwater acidity is probably causing the decline in fish and salamander reproduction. This acid rain is a by-product of air pollution.

A number of studies have shown that the main sources of this "acid rain" are sulphur and nitrogen oxides. Sulphur oxides are by-products of the burning of fossil fuels. Nitrogen oxides are by-products of gasoline burning. These sulphur and nitrogen oxides combine with atmospheric moisture to form nitric acid and sulphuric acid. Then the pH of normally slightly acidic rainfall drops dramatically. Rainwater pH of 3.8 and lower have been recorded in various parts of the United States. The low pH values present threats to plant life as well as to aquatic life forms.

Several alternatives to the problem of acid rain exist and are listed below.

1. Add lime, a buffering agent, to the lakes.

2. Allow the lakes to become "dead lakes."

3. Set new, tougher international pollution laws and standards.

4. Significantly reduce fossil fuel consumption.

5. Develop acid-tolerant fish, as some universities have been working on.

6. Close down industries which are major contributors to nitrogen and sulphur oxides.

7. Do nothing at all.

In small groups, discuss advantages and disadvantages of the alternatives listed above. Make a list of social, moral, and legal questions which might arise from the acid rain situation.

Land Ownership and Land Use: A Question of Values and Rights

Note to the teacher: This activity is divided into three sections. The first section might be appropriate to develop within a unit on maps and mapping, or a unit on soils or pollution. The first section puts the individual students in the position of a landowner whose property is affected by land uses adjacent to his property. In this context, students and other people often recognize the limitations of the rights of people to use their land in any way they see fit.

Make sure you discuss the reasons and the reactions of the students to the first section before they proceed to the second section. Preferably, separate the sections on different pages which are handed out one at a time.

The second section places the students in the situation of choosing between their economic self-interest and a potential abuse of their land. It also puts them in the position of choosing between a regional pollution situation and the potential abuse or decline in value of their property. Here, the situations posed become more like those encountered in real life. The need of society and the rights of the individual that come into conflict must be balanced.

The last section asks the students to resolve their feelings and beliefs about their individual rights as property owners and the needs and values of the society. This section will be difficult for students who have really reflected upon the nature of the conflict posed by the preceding two sections. Discussion following this section is essential, and teachers should be careful to respect the views of students, while continuing to point out situations in which conflicts will still have to be resolved.

Section A:

Ownership of private property and land in the United States is a right so basic that we all take it for granted. Most people feel that if they own land and pay taxes on it, they should have the right to use it as they see fit. Below are some situations in which individual landowners are using their land as they wish. After each situation, indicate whether you agree, disagree, or are unsure about whether or not they should be allowed to use their land in this particular way BY CIRCLING ONE OF THE THREE CHOICES.

In making your decisions on each situation, assume that you own land right next to the land in the situation described, and that you have your home on that land, which is a beautiful, tree-covered twenty acres just like your neighbor's land.

1. An individual owning the land next to your land has just signed a contract with a large nearby city to use the land for the city's sanitary landfill. All the city's garbage will be dumped in this landfill.

You agree are unsure disagree with this use.

2. The person owning the land next to yours has just signed an oil lease on the property with a major oil company who plans to drill for oil on that land. To drill effectively, all the timber on the land will be cleared and at least twenty drilling rigs will be set up on this small parcel.

You agree are unsure disagree with this use.

3. The person owning the land has decided to develop a dairy cattle feed lot on the land. This feed lot will have over 1000 head of cattle on it at all times.

You agree are unsure disagree with this use.

4. The person owning the land next to yours plans to develop a shopping center on it. Most of the property he owns will be covered with stores and an asphalt parking lot.

You agree are unsure disagree with this use.

Indicate your reason(s) for your choices in each of the four situations on the numbered lines below.

1. ______________________________

2. ______________________________

3. ______________________________

4. ______________________________

Section B:

Now, answer the following questions, giving your reasons for your answers.

1. Would you make a different decision about situation number 1 if the alternative to the landfill site was to continue dumping the city's garbage into the ocean nearby?

____

2. Would you make a different decision about situation number 2 if the oil company expressed an interest in buying your twenty acres at a price of $5,000 per acre?

____

3. Would you make a different decision about situation number 3 if the neighbor offered to supply you with free beef for as long as you owned your land?

____

4. Would you make a different decision about situation number 4 if you knew that the building of the shopping center would supply 250 new jobs to your community, thus helping many unemployed people in the area find work?

____

Discuss your choices and decisions about each of the situations in small groups and then complete the following sentences:

Section C:

1. A person should be free to use his or her land

____

____

____

2. If I buy land and have the opportunity to make a profit on it by using it in a certain way, I should

____

____

____

3. With regard to land ownership, the government should

4. Land is (circle one and complete the sentence) more less important than the person's right to own it because

Astronomy and Astrology

Note to the teacher: This activity may initiate a number of related discussions. Allow ample time for students to discuss their feelings and experiences on these issues.

Since the popularity of the song "The Age of Aquarius," from the Broadway musical "Hair," interest in astrology has increased dramatically. People ask each other, "What's your sign?" Radio stations and newspapers regularly present horoscope programs and columns.

1. What is the difference between astronomy and astrology?

2. Do you ever read or listen to horoscopes? Has your horoscope ever come true?

3. What is your astrological sign? (Cancer, Leo, etc.)

4. A group of leading scientists in 1975 denounced astrology and said it was ridiculous to believe that the position of the stars and planets governs a person's fate. They said fate is determined by environment, education, heredity, and other factors, but not the stars. What do you think about the statement of these scientists?

5. What are some possible consequences of believing in astrology?

6. What are some possible consequences of not believing in astrology?

Families and "Life-style"

Note to the teacher: Ask your students to respond to the questions in Part I. Tally and record the responses in a table like the one below, so that the responses are visible to the entire class. You may also want to graph the data and find the mean for each generation. (It is not important that you get totally accurate data. Your data only needs to demonstrate a trend.)

Part I

1. How many children were in your mother's and father's family?
2. How many children are in your present family?
3. How many children do you think you would like to have?

Number of Children in Family

	10-8		7-5		4-3		2 or less	
	Mother	Father	Mother	Father	Mother	Father	Mother	Father
Parent's Family								
Your Family								
Your Future Family								

Part II

Ask your students to offer possible explanations for the type of data that was compiled in the above table. See if any students are willing to share their views regarding their responses to question 3 in Part I.

Part III

On a piece of paper, have your students answer the following:

1. What type of vocation do you plan to pursue?
2. What are the major components of the type of "life-style" you feel you would cherish for most of your life?
3. Do you feel your vocation and lifestyle are compatible with the number of children you feel you would like to have? Explain your answer.

Genetic Counseling

Note to the teacher: This activity is designed to inform your students about genetic counseling and allow them to consider their feelings about this service and some possible social problems that it could cause in the future. Some positive and negative effects of genetic counseling are listed below:

Pros

1. Couples who have a high risk for passing on potential genetic disease should be informed of these chances.

2. Couples that are informed of their chances of having a child with a specific genetic disease are able to psychologically prepare themselves for this possibility.

3. Genetic counseling can assist in preventing the occurrence of birth defects that are due to genetic inheritance.

Cons

1. How "high" is a high risk for genetic disease?

2. Some couples may not understand the information conveyed to them by a genetic counselor.

3. If a counselor is given the right to be prescriptive, some individual rights will be taken away.

4. Can a couple really psychologically prepare themselves for the problems of raising a child that is afflicted with a serious genetic disease?

If your students are not familiar with genetic screening, we suggest you precede this activity with the one entitled "Genetic Screening."

Genetic counseling provides and interprets medical information based on expanding knowledge of human genetics. Its major goal is to prevent the occurrence of birth defects that are due to genetic inheritance.

Although individual birth defects may seem to be relatively infrequent--occurring in about 7 per cent of all births--each year about 250,000 American babies are born with physical or mental defects of varying severity.[1]

Using the basic laws of genetics, a genetic counselor can predict the probability of recurrence of a given abnormality in the same family. For example, Tay-Sachs disease is inherited by a recessive gene. A child afflicted with this disease begins to show signs of central nervous system degeneration at about six months. These symptoms continue to increase, usually ending with death before the child's fourth birthday.[2]

If both parents are carriers of Tay-Sachs, they will have a 25 per cent chance of having a Tay-Sachs child with each pregnancy. The following diagram illustrates this:

Symbols:
T = normal gene
t = Tay-Sachs Gene

Chart Interpretation:
TT 25% both normal genes
Tt 50% carriers
tt 25% Tay-Sachs

♂ Tt — Carrier Parents — ♀ Tt

Possible Offspring:

	T	t
T	TT	Tt
t	Tt	tt

Please respond to the following statements and questions. Time will be alotted to discuss your responses.

1. Genetic counseling will benefit mankind.

   Agree _______
   No Opinion _______
   Disagree _______

2. I feel that, at some time, I would consider using the services of a genetic counselor.

   Agree _______
   No Opinion _______
   Disagree _______

3. Genetic counseling should be a free service to all members of our society.

   Agree ______
   No Opinion ______
   Disagree ______

4. Students in high school should be made aware that genetic counseling exists.

   Agree ______
   No Opinion ______
   Disagree ______

5. The government should provide adequate publicity of genetic counseling on TV and radio, so more adults would be aware of this service.

   Agree ______
   No Opinion ______
   Disagree ______

6. What is the difference between genetic counseling and genetic screening?

7. Should a genetic counselor have the legal right to prescribe medical treatment (i.e., sterilization, abortion) for a couple known to be carriers of a genetic disease? Explain your answer.

8. Do you feel the government should require couples that have a child with a genetic disease and are intending to have more children to seek the services of a genetic counselor? Explain your answer.

REFERENCES

[1]Genetic Counseling, booklet prepared by the National Foundation for the March of Dimes.

[2]Stine, Gerald James, *Biosocial Genetics: Human Heredity and Social Issues,* Macmillan Publishing Co., Inc., New York, New York, 1977.

Genetic Screening

Note to the teacher: This activity has two distinct purposes: to inform your students about genetic screening and to involve them in considering some social problems that this technique could pose to society. Some pros and cons of this process that may be discussed are:

Pros

1. Genetic screening can identify carriers of a specific genetic disease and through proper counseling the carriers can be informed of the possibility of passing the trait on to future offspring.

2. Genetic screening can identify a person with a specific genetic disease. If treatment is available, the person can be given aid.

3. Mass governmental screening programs could provide this service to everyone.

Cons

1. Mass screening is a beginning of increased governmental involvement in formulating genetic criteria for human reproduction. This could lead to other legislation calling for mandatory abortion or sterilization, in an attempt to preserve the human "gene pool."

2. There have been cases of discrimination of specific carriers of genetic diseases. (i.e., Some airlines have refused to hire black stewardesses who are carriers of the sickle cell gene. Life insurance companies have refused coverage to carriers of a specific genetic trait or placed them on a high risk.)

3. It may seem economically unwise to screen everyone for a specific genetic trait. A possible alternative could be to screen specific groups that dominantly carry a certain disease.

4. Some argue that certain mass screening programs have been developed in too much haste. (i.e., PKU: there have been cases of normal children misdiagnosed. These children can be harmed as much as a PKU child that is not placed on a special diet.)

Be sure to allow enough time for your students to present their views regarding genetic screening.

Genetic screening is a general term for the process of determining carriers of a specific genetic disease. Many different tests are used to identify the carriers of specific genetic traits. For example, sickle cell anemia is a disease caused by a co-dominant gene and is most common in blacks. A person inheriting two genes that produce sickle shaped red blood cells will usually die before age 30.[1] The only way that this disease can be inherited is from parents who are both carriers of one sickle cell gene. (This means that they carry one gene for normal red blood cells and one for sickle shaped red blood cells. Because both genes are expressed, a carrier will have the potential to produce both normal and sickle shaped red blood cells. Therefore, a carrier can be identified by analyzing his/her red blood cells.) The following diagram illustrates how this gene is transmitted.

Symbols:
N = normal red blood cells
S = sickle cell gene

Chart Interpretation:
NN 25% both normal red blood cell genes
NS 50% carriers
SS 25% sickle cell anemia

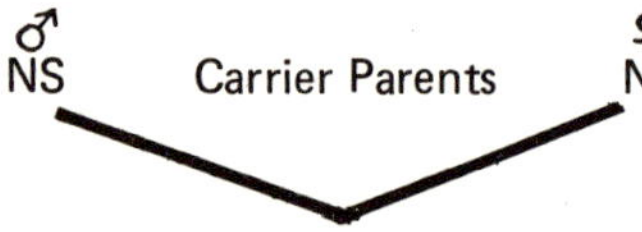

Possible Offspring:

	N	S
N	NN	NS
S	NS	SS

Nine states in the United States plus the District of Columbia have mandatory screening programs to determine carriers of the sickle cell gene. In Kentucky, a fine is assessed if any person fails to be screened for sickle cell anemia when he/she is applying for a marriage license. This law is to make people aware of the presence of this gene and not intended to prohibit people from getting married.[2]

Please respond to the following:

1. Do you feel that the government should pass laws for mandatory genetic screening programs? Explain your answer.

2. Can you think of any social stigma(s) that could be placed on individuals who are found to be carriers of a specific genetic trait? List these and explain your rationale.

3. What possible social implications could mass screening pose to our society?

4. Phenylketonuria (PKU) occurs in about 1 out of 10,000 to 20,000 births in the United States. Do you feel that this disease has too low a frequency to spend money on a mass screening program? Explain your answer.

5. Indicate your feelings regarding genetic screening by checking the appropriate spaces below:

   The government should pass laws supporting mandatory genetic screening programs.

   Agree____
   No Opinion____
   Disagree____

   The government should support voluntary genetic screening programs.

   Agree____
   No Opinion____
   Disagree____

   The government should fund educational efforts to inform the public about genetic screening.

   Agree____
   No Opinion____
   Disagree____

REFERENCES

[1]Stine, Gerald James, Biosocial Genetics: Human Heredity and Social Issues, Macmillan Publishing Co., Inc., New York, New York, 1977.

[2]Ibid.

Amniocentesis

Note to the teacher: This lesson is intended to allow your students to consider some of the moral and ethical problems involved with the process of amniocentesis. Some of these could be:

1. Should an amniotic tap be performed if a couple is 100 percent against having an abortion?

2. Should an amniotic tap be allowed only to determine the sex of the unborn? (In a situation where there is no suspicion of a sex-linked disease being present in either of the parents' genes.)

3. Will the government develop legislation to call for mandatory amniotic taps in the cases of couples known to be carriers of a specific genetic disease?

4. Will amniocentesis lead man away from attempting to find treatment for specific genetic disorders? It may prove to be more convenient to prevent defective individuals rather than provide treatment for them.

We suggest that you provide time for your students to share their answers and comments with their classmates.

John and Mary Smith's first child seemed normal at birth, but later he proved to be the victim of a serious birth defect. He was physically handicapped and mentally retarded.

The Smiths did their best for him, but they yearned for a normal, healthy child. Their friends and parents urged them to try for another child. They told John and Mary that the chances of their having another handicapped child were extremely low.

Finally, it looked as though the Smiths' wish was going to be fulfilled. Mary was pregnant and everything seemed to be going fine.

However, today John and Mary were told by a doctor that there is a technique called amniocentesis that may be able to determine whether the fetus will develop into a healthy child. In this process, a hypodermic needle is inserted through the abdominal wall and into the amnion. A small portion of the amniotic fluid is withdrawn and examined. This fluid contains some of the same cells that comprise the developing embryo or fetus. The doctor showed them the following diagram describing the process of amniocentesis.

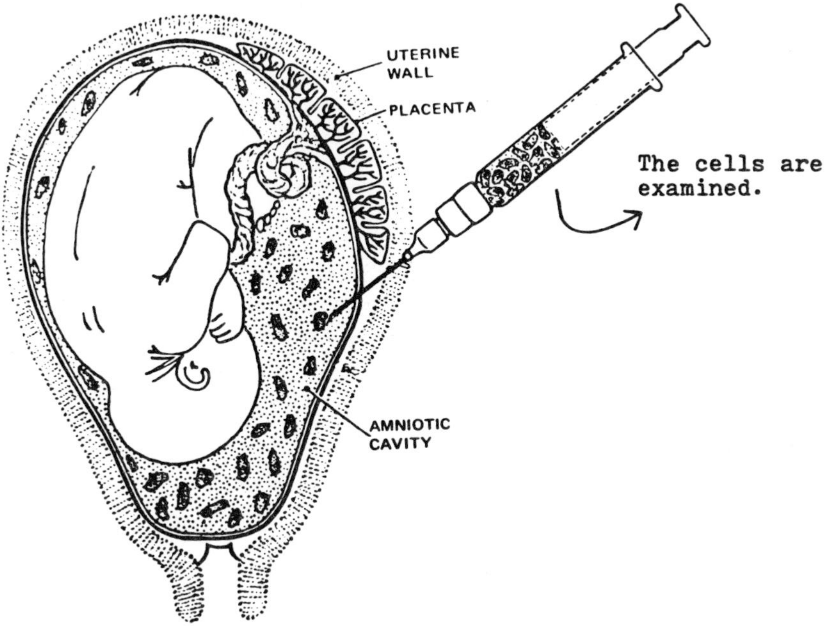

In addition, the doctor informed them that data compiled from 1971-75 shows amniocentesis to be a low risk to either the mother or the unborn.[1] However, the information obtained from this technique may lead to some difficult decisions.

Please respond to the following:

1. What possible moral and ethical problems could this technique raise for John and Mary Smith?

2. If you were John Smith, would you want Mary to have her potential child examined for any defects? Explain your answer.

3. What are some moral and ethical issues that this technique could pose to our society in the future?

REFERENCES

[1]Stine, Gerald James, Biosocial Genetics: Human Heredity and Social Issues, Macmillan Publishing Co., Inc., New York, New York, 1977.

Earthquakes and the Use of Scientific Knowledge

Note to the Teacher:

This three-part activity is designed to (1) raise the level of awareness of students of earthquake hazards and people's involvement with these hazards, and (2) pose for students the problem of the use of scientific knowledge to deal with everyday problems.

Part A asks students to respond to questions involving science and governmental activities and support of research. In effect, this part sets the stage for the subsequent problems involving the earthquake prediction and control situations.

Part B is a role-playing simulation of the kind of problem which is increasingly being faced by U.S.G.S. staff working in earthquake prone areas of the United States. The scenario presented in Part B is not unrealistic.

Part C introduces the problem of the use of new and unproven solutions to problems which threaten great human loss and suffering if not solved. Although the technique described is fabricated, the problem is a valid one and involves a variety of spin-off consequences depending upon the decisions made. Be sure to discuss the consequences of either decision in Part C with your students.

PART A:

Earthquakes are among the most potentially dangerous natural phenomena which occur on Earth. Recently, much effort and money has been spent on research which may allow us to predict the occurrence of an earthquake before the fact. Other researchers are attempting to determine whether or not earthquakes can be controlled or even prevented by drilling wells near a "locked fault" and injecting fluids into the rock in the region. The theory is that these fluids may lubricate the fault, allowing minor earthquakes to occur, thus relieving the pressures and stresses which would otherwise build up and suddenly release in the form of a major earthquake.

Below are a number of statements which are related to the choices we might make about such techniques. Indicate in the area next to each item whether you: strongly agree (SA), agree (A), are unsure (U), disagree (D), or strongly disagree (SD) with each statement.

------- 1. The federal government should spend all the money necessary to develop a fool-proof method of earthquake prediction and control.

——— 2. Experiments such as those described above should be conducted at every available fault site where there is a chance of a major earthquake.

——— 3. If we can predict earthquakes, we can prevent large losses of property and life.

——— 4. Scientific research should be funded on the basis of the extent to which it will save human lives.

——— 5. Scientific research which has no direct benefits to people should not be supported by federal tax dollars.

——— 6. When scientific research allows us to predict an earthquake with 50% accuracy, we should immediately begin to use the research to save human lives.

——— 7. It is wrong to have information which could possibly save human lives and not use it.

——— 8. If real experiments on potential faults which might release a major earthquake are necessary in order to learn how to control earthquakes, it would be wrong not to conduct them.

PART B:

Your group is composed of the administrators and head geologists of the United States Geological Survey. You are in a high level meeting because of some disturbing information which has just been collected by your staff at U.S.G.S.

Recent research conducted by the Survey has just been distributed within the Survey. It has not been released to the news media yet. The research indicates that there is a 50% chance that a major earthquake of Richter 8.3 magnitude will occur in the San Francisco area within the next five days. There is a 75% chance that the earthquake will occur in thirty days.

Your group is faced with making the decision of what to do with this information. Should you release the information to the news media or not?

Discuss the situation and reach a consensus. Be prepared to explain why you made the decision you did.

PART C:

Hold everything! The head geologist on your earthquake control project has just come in! She is excited by this information which is causing you so much trouble. Her team has just completed a small experiment in Colorado where they were successful in releasing the pressure along a fault with surface explosives. By timing the explosions carefully, the team has been able to cause minor earthquakes along the fault and thus relieve the pressure on it. Her estimate is that there is only a 25% chance that these explosions would trigger the major quake and about a 75% chance that they would be able to release the dangerous pressure along the San Francisco fault with minor quakes.

The problem is that the area around the fault will be damaged by the surface explosions, and as a result, about 10 million dollars of damage will result. Should your group allow her to attempt to release the pressure?

Discuss the situation, reach a consensus, and be prepared to explain your decision and justify it.

Human Population and World Food

Note to the teacher: Ask your students to read the informational material in this activity. Then, allow them to proceed through Parts I and II with ample time for small and large group discussion.

"The Generations pouring
From time of endless date,
In their going, in their flowing,
Ever form the steadfast state;
And humanity is growing
Toward the fullness of her fate."

—Herman Melville (1819-1891)

During the nineteenth century words like those above were probably ignored by their readers. However, in the twentieth century many people look at world population as the most significant environment problem facing mankind. The earth has a limited amount of space and resources. These limiting factors directly affect the quantity of food that can be produced.

It has been predicted that the world population will double in 35 years.[1] This will raise the world population from four to eight billion and create an even greater world food problem than we now face. Some biologists believe that not everybody will be able to survive in the future. They argue that soon it will be necessary to decide who is to be saved and who will be allowed to starve.

Three alternatives have been offered as solutions to the world food problem. They are as follows:

1. Send food to only those countries that have developed a plan to limit population growth.

2. Decide who will or will not starve by a system called *triage.* Triage is the method used in large medical disasters, most often in battle-field hospitals, to determine who is to be treated when doctors and medicine are in short supply. Patients are separated into three groups: It is only this last group who needs treatment to survive who receives medical attention. The other two groups are left to die or get well.[1]

TRIAGE

GROUP I	GROUP II	GROUP III
People who will die even if treated.	People who will survive with or without treatment.	People who will survive only if they are treated.

Triage appears to be extremely harsh treatment to starving nations; however, some biologists feel that because there is a limited supply of food that can be grown on this earth, aid should be sent to only the countries that will benefit the most.

3. Launch a worldwide educational campaign to educate people in the area of population control and new food-production techniques.

Part I

In groups of four or five, discuss which of the above alternatives seems to be the best solution to an apparent world population and food problem.

Part II

In the same groups, discuss other possible alternatives to the world population and food problems than those cited in this activity.

REFERENCES

[1] Ehrlich, Paul R., and Ehrlich, Anne H., Population Resources Environment, W. H. Freeman and Co., San Francisco, California, 1972.

[2] Hilton, Bruce, The Ethics of Population Control: An Overburdened Earth? Science and Mankind, Inc., White Plains, New York 10603, 1975.

A Totally Implantable Artificial Heart

Note to the teacher: Challenge your students to role play (in groups of four or five) the activities of the panel organized by the Heart and Lung Institute. Ask them to list some of the moral and ethical problems the totally implantable artificial heart (TIAH) will pose to our society. Allow time for each group to share their views with the rest of the class.

Some possible moral and ethical questions that the TIAH could pose to our society are as follows:

1. Will the definition of death need to be modified for people who are TIAH recipients?

2. If Plutonium 238 is used as a source to power the TIAH, what possible radiation effects could this have on the recipient and those close to him or her?

3. What method will be used to determine who receives the TIAH? Will only rich people be able to afford it?

4. What effect will the TIAH have on our population?

In July, 1972, a panel composed of three physicians, two economists, two lawyers, one sociologist, one ethicist, and one political scientist met to discuss the moral and ethical problems involved in the use of a totally implantable artificial heart (TIAH). This panel was organized by the Heart and Lung Institute, who has developed the TIAH.

The TIAH is a blood pump constructed of synthetic materials driven by a motor and power source. It is designed to be permanently implanted in the human body. Three alternate power sources are being considered for the TIAH: (1) an electric motor powered by a biological fuel cell; (2) an electric motor powered by rechargeable batteries; and (3) a nuclear powered engine using Plutonium 238.

Presently, the Heart and Lung Institute has conducted extensive animal experimentation on TIAH's utilization of these three energy sources.

It is estimated that 17,000 to 50,000 people could benefit from the TIAH annually. However, the use of the TIAH presents some serious moral and ethical problems in our society.*

*Reprinted from THE AMERICAN BIOLOGY TEACHER, Volume 40, No. 2, February 1978; courtesy of the National Association of Biology Teachers, Inc., 11250 Roger Bacon Drive, Reston, Virginia 22090.

High Blood Pressure and Life Style

Note to the teacher: This is a two part activity. It is intended to provide your students with information regarding high blood pressure. Along with this information, your students will be given the opportunity to examine certain components of their life-style. Hopefully, this experience will assist them in identifying whether or not their present life style could lead to the development of high blood pressure.

The "Life-Style Questionnaire" is intended to be for your students' personal benefit. If any student elects to keep his/her answers confidential, this choice should be respected. However, a discussion regarding the factors contributing to high blood pressure could be valuable for your students. You may want to discuss the government's role in informing people about the relationship between high blood pressure, coronary heart disease, and life style.

It may also be interesting to pursue the following question: Should the federal government shift support from developing a totally implantable artificial heart to a mass public educational effort to inform people as to the possible prevention of high blood pressure and other heart disease and to the importance of having blood pressure checked regularly? (We recommend that you use the activity entitled, "A Totally Implantable Artificial Heart," before posing this question to your students.)

Part I: Distribute this questionnaire to your students.

Life-Style Questionnaire

This questionnaire is designed to help you evaluate some specific components of your life-style. Your answers are completely confidential and you need only share them with your peers and teacher if you so desire.

I. Listed below are some common foods. Indicate your preference by placing a check by either A or B. (If you do not like either choice, leave it blank.)

1. ------ A. Fish
   ------ B. Beef Steak

2. ------ A. Shrimp
   ------ B. Lamb Chop

3. ------ A. Chicken Breast
   ------ B. Pork Chop

4.	______	A.	Butter
	______	B.	Margarine
5.	______	A.	Skim Milk
	______	B.	Whole Milk
6.	______	A.	Vanilla Yogurt
	______	B.	Vanilla Ice Cream
7.	______	A.	Ladyfinger Cookies
	______	B.	Brownies
8.	______	A.	Rice Pudding with Raisins
	______	B.	Bread Pudding with Raisins
9.	______	A.	Cheddar Cheese
	______	B.	Cottage Cheese
10.	______	A.	Whipping Cream
	______	B.	Half & Half

Please respond to the following questions by checking the blank corresponding to either yes or no.

		Yes	No
1.	Do you smoke cigarettes?	______	______
2.	Do you smoke a pipe or cigars?	______	______
3.	Do you exercise regularly?	______	______
4.	Do you feel that you receive an adequate amount of rest each day?	______	______
5.	Do you feel that you are overweight? (About 15-20 pounds overweight for your height and frame.)	______	______
6.	Do you like to use a lot of salt to flavor your food?	______	______
7.	Do you feel that your life is filled with "high stress" situations?	______	______

Part II: After your students have completed the questionnaire, distribute this information regarding high blood pressure.

High Blood Pressure

Uncontrolled high blood pressure (hypertension) is the primary cause of 60,000 deaths each year in the United States.[1] In addition, high blood pressure is a contributing factor to heart attacks and strokes. It is estimated that 23 million Americans have high blood pressure.[2] Among American blacks, the prevalence of this disorder is twice as high as among whites.[3] The reason for this is not completely understood.

High blood pressure has been most common in late middle age in the United States. However, in the last few years, this disorder is becoming more common among younger people.

Physicians generally agree that the following items contribute to the development of high blood pressure.[4]

1. lack of exercise
2. inadequate rest
3. being overweight
4. high salt diet
5. smoking (cigarettes being the worst offender)
6. high serum cholesterol
7. high stress in everyday life

Physicians also feel that if many people would avoid these seven items, through a change in life-style, a significant number of cases of high blood pressure would decrease annually.

Now, compare your answers on the "Life-Style Questionnaire" to the following information.

1. Food Preference, The cholesterol data were adapted from these sources: Hursh, L. M., Coronary Heart Disease: Risk Factors and the Diet Debate, National Dairy Council, Chicago, Illinois, 1975.
   Mayer, Jean, A Diet For Living, Pocket Books, New York, New York, 1977.

		Amount of Food (grams)	Amount of Cholesterol (milligrams)
1.	Fish	85	34-75
	Beef Steak	85	80

2.	Shrimp	85	128
	Lamb Chop	85	83
3.	Chicken Breast	85	63
	Pork Chops	85	76
4.	Butter	5	12
	Margarine	5	0
5.	Skim Milk	236	5
	Whole Milk	236	34
6.	Vanilla Yogurt	227	17
	Vanilla Ice Cream	227	35
7.	Ladyfinger Cookies	44	157
	Brownies	44	45
8.	Rice Pudding with Raisins	133	15
	Bread Pudding with Raisins	133	85
9.	Cheddar Cheese	28	28
	Cottage Cheese	28	4
10.	Whipping Cream	15	20
	Half & Half	15	6

According to this chart, do you prefer foods that are generally lower or higher in cholesterol? What effect would your preference have on your serum cholesterol?

Now, compare your answers in Part II of the questionnaire to these items. In addition to high serum cholesterol, each is also considered to be a contributing factor to high blood pressure.

1. smoking
2. lack of exercise
3. inadequate rest
4. being overweight
5. high salt diet
6. high stress in everyday life

As evidenced by your responses on the questionnaire, do you feel that you run a high, medium, or low risk of developing high blood pressure? Explain your answer.

Please complete the following unfinished sentence:

This activity has assisted me in identifying

REFERENCES

[1]Guidelines for the Evaluation and Management of the Hypertensive Patient, DHEW Publication No. (NIH) 76-744, September 1, 1973.

[2]Ibid.

[3]Hursh, L. M., Coronary Heart Disease: Risk Factors and the Diet Debate, National Dairy Council Publication, Chicago, 1974.

[4]Guidelines for the Evaluation and Management of the Hypertensive Patient, DHEW Publication No. (NIH) 76-744, September 1, 1973.

DO YOU AGREE OR DISAGREE?

Note to the teacher: This activity could be used as an introduction to a unit on drugs. Some common controversial topics are raised regarding drugs and drug usage. We suggest that you allow ample time for discussion after the students respond to the following statements.

1. Read the following statements. Circle the response that best fits your feeling concerning each particular remark. The responses are as follows:

SA	Strongly Agree
AS	Agree Somewhat
NO	No Opinion
DS	Disagree Somewhat
SD	Strongly Disagree

1.	Marijuana should be legalized.	SA	AS	NO	DS	SD
2.	Drugs are analogous to alcohol.	SA	AS	NO	DS	SD
3.	Police should crack down on the drug pusher more than the user.	SA	AS	NO	DS	SD
4.	Most drugs can be beneficial to man if used for medical purposes.	SA	AS	NO	DS	SD
5.	The use of marijuana leads to the use of "harder drugs."	SA	AS	NO	DS	SD
6.	Drug users are emotionally maladjusted people.	SA	AS	NO	DS	SD

7. Drug users are mainly from families of low economic status.	SA	AS	NO	DS	SD
8. Most drug users resort to crime to support their habit.	SA	AS	NO	DS	SD
9. Society should treat the drug user like a sick person rather than a criminal.	SA	AS	NO	DS	SD
10. Drug education should be part of every school system's curriculum.	SA	AS	NO	DS	SD
11. Each person should have the right to decide if he wants to use marijuana or drugs rather than have his government decide this for him.	SA	AS	NO	DS	SD

II. Now, select two of the above statements and in a paragraph explain why you responded the way you did.

Uniform Anatomical Gift Act

Less than a few decades ago, the miracle of organ transplantation was beyond belief. Today, medical research has significantly improved the technique of kidney, cornea, and heart transplants. However, there are still great shortages of available organs.

Most states have passed a Uniform Anatomical Gift Act or similar laws. The purpose of these laws is to make it possible for anyone who desires to give one or more of his/her organs to some other individual to do so by simply signing a uniform donor card in the presence of two witnesses. An example of a uniform donor card is shown in Figure 1.

FIGURE 1

UNIFORM DONOR CARD

of ______________________________

Print or type name of donor

In the hope that I may help others, I hereby make this anatomical gift, if medically acceptable, to take effect upon my death. The words and marks below indicate my desires.

I give: (a) — any needed organs or parts
(b) — only the following organs or parts

____

Specify the organ(s) or part(s)

for the purposes of transplantation, therapy, medical research or education:

(c) — my body for anatomical study if needed.

Limitations or
special wishes if any: ______________________________

Signed by the donor and the following two witnesses in the presence of each others:

____

Signature of Donor — Date of Birth of Donor

____

Date Signed — City & State

____

Witness — Witness

This is a legal document under the Uniform Anatomical Gift Act or similar laws.

For further information consult your physician or

KIDNEY FOUNDATION OF IOWA
8611 Hickman Road, Des Moines, Iowa 50322

Uniform Anatomical Gift Act

Indicate your feelings toward the following statements by placing the letter that precedes each statement on this continuum.

AGREE	AGREE SOMEWHAT	NO OPINION	DISAGREE SOMEWHAT	DISAGREE

A. Organ transplantation is beneficial to humans.

B. At this time, I feel that I might consider signing a uniform donor card.

C. I feel uneasy about donating some parts of my body when I die.

D. There could be some people to whom I would not want to donate organs.

E. If I needed an organ transplant, I would be in favor of receiving one.

Note to the teacher: We suggest that you discuss these statements with your students. We found it useful to tally student responses to each statement. This could demonstrate some inconsistent thought patterns in the minds of your students regarding organ transplantation. For example, if a person agrees with statement E, his thought pattern would be inconsistent if he disagreed with item B.

If you are using this activity with students who are not legal adults, you may want to inform them that they may carry a uniform donor card if they have it countersigned by their parents or legal guardians.

A Matter of Life and Death

Note to the teacher: This is a two part activity. Be sure to allow enough time for discussion after each segment. Additional questions that could be probed at the end of this activity are:

1. Do you feel that a panel of doctors should decide who receives specific transplants? Why or why not?

2. Would a panel of different professionals (lawyers, clergy, etc.) be a better alternative than having a panel of medical doctors only? Explain your answer.

3. Do you feel a "first come, first served" policy would be better in the case of determining organ transplant recipients? Explain your answer.

Part I

In groups of four or five, you are to assume the role of a panel of doctors, all specialists in heart surgery. The topic of your discussion is to select one of the three people, each of whom will die without a heart transplant, to receive the one available heart. Each potential recipient is compatible with the protein of the donor.

The panel of doctors has only 15 minutes to make their decision from the following people:

1. A 65-year-old male medical doctor
2. A 30-year-old mother of two children
3. A 17-year-old high school senior

At the end of the 15 minutes, your group will be asked to explain the rationale behind your decision. Your group may select a spokesperson to present the rationale.

Part II

Hold Everything!

The following information has just arrived for the members of the panel. In the same groups, discuss this information to determine if it will change your last decision. Again, you will be given fifteen minutes for consultation. Then you will be asked to share your decision and rationale with the rest of your classmates.

1. The 65-year-old male medical doctor has been ailing for the last five years. Research that he conducted seven years ago has been instrumental in a recent successful chemotherapy for leukemia. However, it is impossible to determine whether this heart transplant will enable him to conduct further cancer research.

2. The 30-year-old mother suffered heart damage three years ago due to secondary effects of an illness. Her chances of leading a normal life after the transplant are about 50-50.

3. The high school senior is a young man who has suffered from a congenital heart ailment since birth. The heart transplant could give this young man a new lease on life. However, there are no predictions as to his chances of total recovery.

Alive or Dead?

Note to the teacher: This activity is divided into four segments. We suggest that you conduct Parts I through III in one class period and allow two periods for segment IV.

The purpose of this lesson is to provide an experience for students to consider the following questions: (1) How does one differentiate between living and non-living? (2) How does one define life and death? (3) Are there certain situations when active euthanasia could be morally "right"?

Part I of this activity deals with the preparation and demonstration of "The Beating Heart." The directions are as follows:

Part I

1. Preparation of "The Beating Heart"

Place a pool of mercury about 3/4" in diameter in a 3-4 inch watch glass. Cover the mercury with 6 M H_2SO_4. Add 1 ml of 0.1 M $K_2Cr_2O_7$. Place an iron nail in the watch glass so that its tip just touches the mercury and the nail lies along a radius of the glass.

2. Demonstration

Conduct this demonstration on an overhead projector so that "The Beating Heart" can be projected on a screen or wall. Slowly add 0.5-2 ml of 18 M H_2SO_4 above the pool of mercury. As soon as rhythmic motion occurs in the mercury, stop adding the acid.

3. Additional Comments

The pool of mercury will continue to beat for hours if the iron is held firmly in place. For this reason, you may prefer to use an iron wire connected to an external handle resting on a desk or table. This allows an easier adjustment of the position of the iron to give maximum beats.

Part II

Note to the teacher: Divide your students into groups of four or five and ask them to discuss the following:

1. Do you feel that the object projected on the screen was alive or dead?

2. Explain the rationale for your answer.

Provide time for each group to share its views with the rest of the class. At the end of this discussion, explain the composition of "The Beating Heart" to your students.

Part III

Note to the teacher: Pass out the following hypothetical situation for your students to read:

John Doe, 18, has been in a coma and connected to a respirator for three weeks. Without the continuous aid of this machine, he would stop breathing. Electroencephalograph readings taken on John Doe, twelve hours apart, indicate almost negligible brain wave activity. However, occasionally (about once every five minutes) John demonstrates the ability to swallow and blink.

John Doe's parents feel that he is dead. They believe that the signs of life exhibited by John are only due to his being connected to the respirator. In addition, having their son connected to this machine is costing the Does over a thousand dollars a week. Their monthly income can in no way pay for this medical service.

The medical staff attending to John have mixed feelings regarding his condition. Some members of the medical team feel that John is still alive. Therefore, in their minds, to remove him from the machine would be an act of homicide.

Tomorrow Mr. and Mrs. Doe are going to court to seek a legal ruling regarding the disconnecting of John from the machine.

Part IV: Role Playing the Court Hearing

Note to the teacher: Divide the class into two groups. One group is to defend Mr. and Mrs. Doe's point of view. They are to prepare a defense for the Does. The other group's assignment is to prepare some statements to show that disconnecting the respirator would be homicide. (Each group may elect one or more spokespeople to present their case. Encourage your students to simulate a court hearing.)

A few questions that you may want your students to consider as they prepare their defense or counterarguments are:

1. How is the John Doe situation similar to "The Beating Heart" demonstration?

2. What is the definition of life and death?

3. Is the present legal definition of death adequate to deal with modern medicine?

You are to role play the judge in this court case. You are to arrive at your decision based on the information presented by each group.

A Definition of Death?

Note to the teacher: (You may want to precede this activity with the one entitled "Alive or Dead?".)

We suggest that you conduct this activity in two segments. It is also important that each group be given enough time to analyze the statutes and possibly formulate their own.

If you would like to inform your students as to which states, on or before February, 1978, had adopted one of the three statutes cited in this activity, the following has been included:

Statute I - Kansas, Maryland, Virginia, New Mexico, Alaska, and Oregon

Statute II - Michigan, West Virginia, Louisiana, Iowa, and Montana

Statute III - California, Georgia, Illinois, Oklahoma, Tennessee, and Idaho

Part I

Please read the following case study:

Mr. James Johnson, 31, was violently assaulted in the parking lot of a local restaurant. He was found by a parking lot attendant in an unconscious condition.

A short time after Mr. Johnson was hospitalized, he regained consciousness. However, his condition was considered critical. He remained in this condition for about one week. Then, his heart-lung functioning stopped and he once again became unconscious. He was immediately connected to a mechanical respirator and has continued in this condition for two weeks. The attending physician has taken a series of electroencephalograph readings on Mr. Johnson ever since his vital signs ceased. The first readings indicated that brain wave activity was present. However, the last three readings, taken twenty-four hours apart, revealed no brain wave activity. Therefore, the physician feels that Mr. Johnson is dead and should be removed from the respirator.

While Mr. Johnson was still conscious, a suspect was taken into custody and charged with his assault. This suspect is aware that if Mr. Johnson is removed from the respirator, he will be charged with second degree murder. Therefore, he and his attorney do not want Mr. Johnson removed from the machine. They argue that the state in which the crime took place has no legal definition for brain death. This state only defines death as the cessation of heart and lung functioning. Consequently, the defendant and his attorney argue that Mr. Johnson is legally alive, and to remove him from the machine would be morally wrong. (Also, the state's action would place the defendant in jeopardy.)

Part II

Because of situations such as the case study you have just read, judges in the United States have requested legislative action to define brain death. Some states have responded to this plea by passing statutes that could be generally grouped into these three categories:[1]

Statute I - Death can be defined by the inability to breathe and by heart cessation. Or, a person can be considered dead if an electroencephalograph (EEG) reading indicates a lack of brain waves.

Statute II - Death is defined on the traditional heart-lung concept. If both organs have stopped, the person is considered dead. However, if a life support system is applied, death can be declared if the attending physician(s) is(are) convinced that the patient has suffered irreversible brain cessation.

Statute III - A third statute is based on a proposal of the Law and Medicine Committee of the American Bar Association. This proposal states that a human body is dead if irreversible brain cessation occurs, as indicated by an EEG.

In groups of four, discuss and analyze the above statutes regarding brain death, recently adopted by some states in the United States. Focus your discussion on these factors:

A. Which statute do you feel you would want your state legislature to adopt? Explain the rationale for your answer.

B. Which statute would you least like to see your state legislature adopt? Why?

C. If you feel that none of the statutes is completely adequate, attempt to devise one that you feel would be more appropriate.

Each group will be given time to share their ideas with the rest of their classmates.

[1] Issacs, Leonard, "Death, Where Is Thy Distinguishing?", Hastings Center Report, 8(1):5-8, Feb. 1978.

The Living Will

Note to the teacher: This activity is designed to initiate discussion about the legal and ethical aspects of a living will. This topic will also encourage discussion regarding the ethical questions surrounding euthanasia. Be sure to allow enough time for all of your students to share their views about the living will issue and any related subjects.

You may want to inform your students that on January 1, 1977, a law went into effect in California regarding the legality of the living will. The law states that the living will:

1. must be renewed every five years.
2. does not apply to pregnant women or comatose patients.
3. only binds a physician to the patient's wishes if it is signed at least 14 days before a person is diagnosed as terminally ill.

Since this law was adopted in California, about six other states have passed similar legislation.

Two weeks ago, Jim Smith was the victim of an auto accident. Since that time, Mr. Smith has remained in a comatose condition and requires a mechanical respirator to sustain his life. Two recent electroëncephalograph readings taken on Mr. Smith, twelve hours apart, indicate almost negligible brain wave activity.

Mrs. Smith wants the attending physician to remove Mr. Smith from the respirator. She contends that her husband, in sound mind, signed a living will. (Figure 1 is an example of the type of will that Mr. Smith signed. It was prepared by the Euthanasia Educational Council, 250 West 57th Street, New York 10019.)

Figure 1

To my family, my physician, my clergyman, my lawyer:

If the time comes when I can no longer take part in decisions for my own future, let this statement stand as the testament of my wishes:

If there is no reasonable expectation of my recovery from physical or mental disability, I, ______________, request that I be allowed to die and not be kept alive by artificial means or heroic measures. Death is as much a reality as birth, growth, maturity and old age--it is the one certainty. I do not fear death as much as I fear the indignity of deterioration, dependence and hopeless pain. I ask that medication be mercifully administered to me for terminal suffering even if it hastens the moment of death.

This request is made after careful consideration. Although this document is not legally binding, you who care for me will, I hope, feel morally bound to follow its mandate. I recognize that it places a heavy burden of responsibility upon you, and it is with the intention of sharing that responsibility and of mitigating any feelings of guilt that this statement is made.

Signed______________________

Date______________________

Witnessed By:______________________

____

____

Mrs. Smith feels that her husband's wishes, indicated by this document, ought to be respected.

Please respond to the following questions and statements. Time will be given to discuss your answers with your colleagues.

1. Put yourself in the position of Mrs. Smith and the attending physician. What effect do you feel that Mr. Smith's living will could have on their decision-making?

2. Do you feel that a living will should be accepted in court as a legal document? Explain your answer.

3. Cite at least two examples of when you feel a person's wishes in a living will should or should not be respected.

Abortion

Note to the teacher: This is an extremely emotional issue. We suggest you employ discretion concerning the use of this activity in your classroom. If you elect to incorporate it in your instruction, we recommend that you allow time for your students to discuss their views. However, do not force anyone to share his or her opinion. An emotional issue of this nature can be an intimidating experience for some individuals.

Abortion is defined as the spontaneous or induced expulsion from the uterus of an embryo or fetus before it has reached a point of development sufficient for its survival (usually up to the 28th week of pregnancy).[1] The ethical nature of this technique has been questioned and debated for centuries.

In 1973, a legal precedence was set in the United States (Roe vs. Wade) regarding abortions performed on pregnant women whose lives were not in danger due to pregnancy.[2] The court ruling allowed a woman the freedom to decide whether or not she wanted to end a pregnancy during the first trimester (about 1-12 weeks). During the second trimester (13th through 24th week), a state may regulate only the abortion procedure in ways related to maternal health. This means that through the first six months of pregnancy, a woman may legally elect to have an abortion for any reason.[3]

While the United States Supreme Court decision had some ethical implications, it did not resolve this issue in the minds of many people. There are still two fundamental ethical questions that are debated:[4] (1) Is the embryo or fetus a "human being," a "person" entitled to life, liberty, and the pursuit of happiness? (2) Does the embryo's or fetus's claimed right to life outweigh the acknowledged rights of the mother?

A. What is your opinion concerning the status of the embryo or fetus? Indicate your feelings on the following continuum by checking the moment you believe human personhood begins in human development.

Conception 1 2 3 4 5 6 7 8 9 Birth

⟶

months from conception to birth

Explain the rationale for your answer.

B. Do you believe a woman's right to the control of her body outweighs the supposed right to life of the unborn? Explain your answer.

C. Next, indicate your feelings toward legalized abortion by checking the appropriate place on the following continuum.

Legalized abortion should be made available to anyone who wishes to end a pregnancy.	Legalized abortion should be made available only in certain situations.	Abortion should not be legalized for any reason.

REFERENCES

[1]Neuman, G., "Abortion," The Encyclopedia of Sexual Behavior, Vol. 1, Hawthorn Books, New York, New York, 1961.

[2]Lockhart, William B., et al., Constitutional Law, 1973 Supplement to 3d Edition, West Publishing Co., St. Paul, Minnesota.

[3]Abortion: A Woman's Right and Guidelines for Counseling, a booklet prepared by the Abortion Counseling Service of Minnesota, 549 Turnpike Road, Golden Valley, Minnesota 55416.

[4]Hunt, Robert, and Arras, John, Ethical Issues in Modern Medicine, Mayfield Publishing Co., Palo Alto, California, 1977.

ESB and You

Note to the teacher: Be sure to allow enough time for your students to share their views regarding ESB with classmates. It may be desirable to have your students discuss their views in small groups before sharing their opinions with the entire class.

Sometimes it can be interesting to look into the future. The following dialogue discribes a technique of the future that could prove to be very beneficial to mankind. The technique is called electronic stimulation of the brain or ESB.

ESB is a direct electrical stimulation of carefully selected parts of the brain with stainless steel electrodes that are implanted surgically. Proper stimulation can evoke different types of emotional and physical responses in a human; for example, it can relieve pain without the aid of drugs and lull you into sleep.

For some physically handicapped people, ESB could cure their disability. For example, ESB could be used by a stroke-paralyzed individual. By using this technique, a stroke victim could stimulate parts of his brain that were not affected by the stroke. This stimulus could possibly move paralyzed parts of the body into coordinated muscular movements. With computerized controls, the victim would have only to push a few buttons in order to instruct his brain electronically as to where, when, and how fast he would like to move. Some other handicapped people it is hoped can be helped by ESB are the blind, the deaf, and those who are prone to epileptic seizures.

It also appears the ESB has potential in regressing patients to early childhood. This could supply psychiatrists with a wealth of information about their patients.

Imagine this possibility: a person properly wired with electrodes might be able to one day plug into a computer at home, dial in the number that corresponds to a particular experience, and then lean back and enjoy it. It is conceivable that by using the right electrodes in the correct sequences, under computerized control, a person might be able to experience just about anything that one can imagine (a trip to outer space, scoring the winning touchdown in the Super Bowl, etc.).

After reading the above explanation of ESB, please respond to the following questions or statements concerning this technique.

1. Do you feel that scientists should continue to probe into the possible uses of ESB? Explain your answer.

2. Make a list of some of the undesirable uses of ESB. What implications could these misuses pose to our society?

3. If you could use ESB to enjoy any human experiences you wished, which ones would you choose? Why would you select these experiences?

REFERENCES

The narrative on ESB has been adapted from the book, As Man Becomes Machine, by David Rorvik, Doubleday and Co., Inc., New York, NY, 1971.

The Cloning of Humans

Note to the teacher: The purpose of this activity is to provide your students with information regarding the technique of cloning and to allow them to discuss some of the moral and legal questions that could arise as a result of this technique. Some questions that your students might list are:

1. Is a clone a legal human being?
2. Should a clone be given the same rights under the United States Constitution as a person now considered to be a legal human being?
3. Who can own a clone?
4. Can a clone be patented?
5. Could man attempt to produce a master race?
6. Who is to decide what are the most desirable human traits?
7. Does a clone possess a soul?
8. Could cloning drastically change our society's concept of motherhood and fatherhood?

Be sure to allow time for each group to share their ideas with the rest of their classmates.

In early March, 1978, it was reported that a scientist had successfully cloned a human.[1] There was scepticism raised in the scientific community regarding the validity of this claim. However, even if the report is false, some scientists do believe that this technique will be used on human cells in the very near future.

Before the above claim, the highest form of animal cells that had been successfully cloned were those of South African clawed frogs (Xenopus laevis).[2] The generalized process of cloning this amphibian is shown in the following diagram.

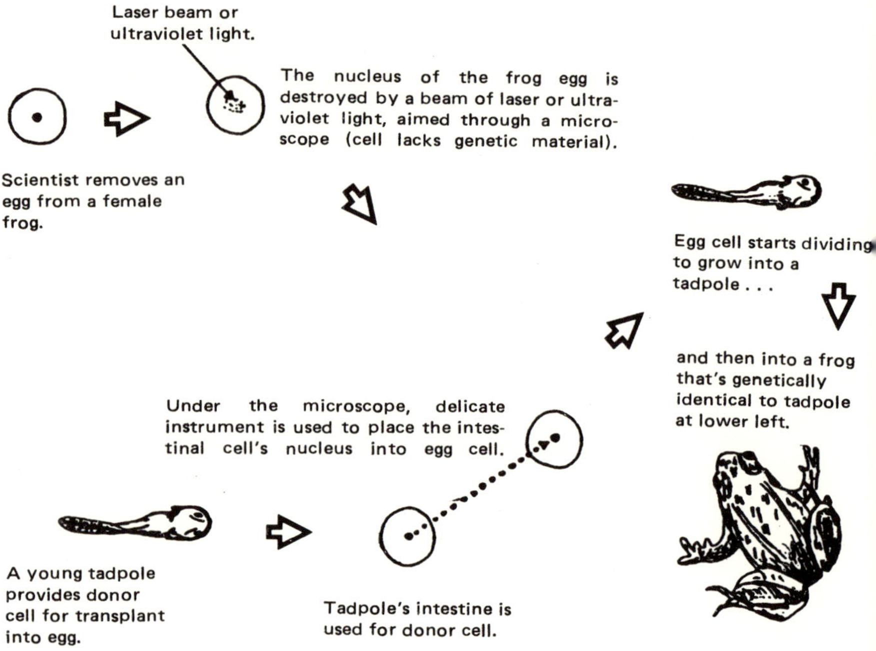

At present, there are no guidelines for the use of this technique. Therefore, a number of questions can be raised regarding its use with humans. In groups of four, examine the following statements and construct a list of some of the moral and legal questions that could result from the cloning of humans.[3] Each group will be given time to share their ideas with the rest of the class.

1. A human birth certificate can only be issued to a person that possesses two biological parents.
2. The United States Patent Appeals Court has ruled that any corporation creating a novel form of life in the laboratory may obtain a patent for the new creation.
3. Humans should only use cloning to reproduce. This process is more predictable than the present means of procreation.
4. Humans could literally replace all life on this earth with "artificial life."
5. "Cloning Banks" could be developed for possible organ transplantation. Clones could be institutionalized and used as a source of spare organs if needed.

REFERENCES

[1]"Suit seeks data on U.S. role in cloning," Duluth News-Tribune, Wed., March 8, 1978.

[2]Stine, Gerald James, Biosocial Genetics: Human Heredity and Social Issues, MacMillan Publishing Co., Inc., New York, New York, 1977.

[3]These five statements were discussed by Jeremy Rifkin on the March 6, 1978, NBC TV Today Show. Rifkin is the co-author of Who Should Play God?, Delacorte Press, New York, New York, 1977.

Offshore Fishing Rights

Note to the teacher: This activity may be appropriate in an earth science class during a study of oceanography. Be sure to have students respond in writing as teams before conducting the discussion.

The small, poor country of Balboa is overpopulated and has 400 miles of ocean shoreline. While poor in most resources, a strong upwelling along the shore of Balboa makes its offshore waters an excellent fishery. Over the past several years, the Balboan government has been attempting to develop a fishing industry to make use of this resource. Recently however, several large countries have sent contingents of their fishing fleets into the Balboan waters and are successfully harvesting large catches beyond the internationally recognized three-mile limit. The Balboan government is angered, claiming that their jurisdiction of the ocean extends out 200 miles and that therefore the foreign fishing is illegal. The foreign fishing enterprise, claim the large countries, is operating legally outside the three-mile limit. Last week, Balboan gunboats seized two foreign fishing vessels and arrested the crews. The crews are scheduled to stand trial in one week for trespassing and theft of Balboan natural resources. The large countries whose crews and vessels have been seized are threatening to take military action if their crews and vessels are not immediately released.

Your team is composed of U.N. representatives from involved countries, including Balboa. You are charged with reaching some compromise which will avert military action and will solve the fishing rights dispute permanently.

Discuss the situation, develop some compromises, and prepare a report to present on this issue to the U.N. General Assembly (your class).

Smoking Versus the Government

Note to the teacher: This activity is designed to allow students to examine the legislative controls that have been placed on cigarette smoking. This specific example illustrates how some state and federal legislation is intended to look after our welfare and provides an opportunity for students to consider to what extent government should enact such laws.

In 1964, the U.S. Surgeon General issued a statement declaring that cigarette smoking was a health hazard. Since this announcement, the federal and a few state governments have enacted legislative controls on the sale and advertisement of cigarettes and the smoking of cigarettes in public places. A summary of the federal and state governmental actions regarding cigarette smoking is listed in Figure 1.[1]

Recently, the Surgeon General has declared that the research conducted from 1964 to present, to determine whether or not smoking was a health hazard, definitely demonstrates that smoking causes death and disease.[2] As a result of these findings, the Secretary of H.E.W. favors more stringent governmental regulations on the cigarette industry.

1. Do you agree or disagree with the Secretary of H.E.W.? Explain the rationale for your answer.

2. Do you feel that any of the governmental actions listed in Figure 1 are unjust? If so, which ones and why?

3. To what extent should the government attempt to pass legislation that is intended to look after our welfare? Explain your answer.

FIGURE 1

Governmental Actions on Smoking

1964	Surgeon General's report on smoking and health determines that cigarette smoking is a health hazard.
1966	Federal Cigarette Labeling and Advertising Act requires, as of January 1, cigarette packs to carry statement: *Caution: Cigarette smoking may be hazardous to your health.*
1967	Federal Communications Commission rules that the "Fairness Doctrine" applies to cigarette advertising, and TV and radio must carry antismoking messages.
1970	Cigarette pack statement changed to: *Warning: The Surgeon General has determined that cigarette smoking is dangerous to your health.*
1971	Radio and TV ads banned as of January 2. Interstate Commerce Commission restricts smoking to rear five rows of interstate buses.
1972	All cigarette advertising must carry the same health warning as on packs, as well as the tar and nicotine content of a cigarette.
1973	Arizona becomes first state to prohibit smoking in all elevators, indoor theaters, libraries, art galleries, museums, concert halls, and buses. All airlines required to designate smoking and no smoking areas in planes.
1975	Minnesota passes Indoor Clean Air Act, which makes smoking illegal in all public places and public meetings except where designated.
1978	Civil Aeronautics Board bans cigar and pipe smoking on all American commercial airlines.

REFERENCES

[1]Oakley, Roy, Drugs, Society and Human Behavior (2nd ed.), The C. V. Mosby Co., St. Louis, Missouri, 1978.

[2]Associated Press, "Smoking More Dangerous Than Supposed in 1964," January 11, 1979.

A BIBLIOGRAPHY OF AUDIO VISUAL/WRITTEN MATERIALS

Films:

"Bertha," Kennedy Foundation - Lowengrad and Brotherhood, 999 Asylum Ave., Hartford, CT 06105, 25 min.

"Evolution by DNA: Changing the Blueprint of Life," Document Associates Inc., 880 3rd Ave., N.Y., NY 10022, 20 min.

"Genetics: Man the Creator," Document Associates Inc., 880 3rd Ave., N.Y., NY 10022, 18 min.

"Man Made Man," CBS Released, McGraw Hill Book Co., Hightstown, NY 08520, 25 min.

"Technology and Values: The New Genetics," UMHE, 2211 E. Kenwood Blvd., Milwaukee, WI 53211, 21 min.

"The Right to Let Die," Kennedy Foundation - Lowengrad and Brotherhood, 999 Asylum Ave., Hartford, CT 06105, 25 min.

"Who Should Survive?" Kennedy Foundation - Lowengrad and Brotherhood, 999 Asylum Ave., Hartford, CT 06105, 25 min.

Slide Tape Presentations: (The following presentations are from Science and Mankind, Inc., Two Holland Ave., White Plains, NY 10603.)

Critical Issues in Science and Society: A World to Feed

Living with Dying

The Ethical Challenge: Four Biomedical Case Studies

The Ethics of Genetic Control

The New Genetics: Rights and Responsibilities

The Science and Ethics of Population Control: An Overburdened Earth

Comprehensive Written Materials:

Ethical Issues in Health Services: A Report and Annotated Bibliography, PB-237-771 National Technical Information Services, U.S. Dept. of Commerce, 1974.

Human Genetics: Information and Educational Materials, DHEW Publication No. (HSA) 79-5132.

Human Genetics: Information and Educational Materials, DHEW Publication No. (HSA) 80-5134.

Readings in Society, Ethics and the Life Sciences, Institute of Society, Ethics and the Life Sciences (Reading Packets 100-4000 and Current Journals), Hastings-on-Hudson, NY 10706.

Reich, T. (Editor-in-chief), Encyclopedia of Bioethics (1st ed.), The Free Press, New York, NY 1978.

Science for Society: A Bibliography (5th ed.), AAAS Commission on Science Education, 1776 Massachusetts Avenue N.W., Washington, DC, 1974.

Science, Technology & Society: A Guide to the Field, National Technical Information Service, U.S. Dept. of Commerce, P.O. Box 1553, Springfield, VA 2216l.

Walters, LeRoy (ed.), Bibliography of Bioethics, Center for Bioethics, Kennedy Institute, Georgetown University, Washington, DC, 1975.

Books:

American Friends Service Committee, Who Shall Live?, Hill and Wang, Inc., New York, NY, 1970.

American Psychological Association, Ethical Principles in the Conduct of Research with Human Participants, Washington, DC, 1973.

Barbour, Ian G., Issues in Science and Religion, Harper Torchbook, New York, NY, 1971.

Beauchamp, T. L., and LeRoy Walters (eds.), Bioethics, Dickenson Publishing Co., Inc., Belmont, CA, 1978.

Beecher, Henry K., Research and the Individual, Little, Brown and Co., Boston, MA, 1970.

Bergsma, Daniel, et al.,(eds.), Ethical, Social, and Legal Dimensions of Screening For Human Genetic Disease, Stratton Intercontinental Medical Book Corp., New York, NY, 1974.

Brim, Orville G., Jr., et al.,The Dying Patient, Russell Sage, New York, NY, 1970.

British Broadcasting Corporation, Morals and Medicine, Oxford Book Co., New York, NY, 1971.

Brody, Howard, Introduction to Ethical Decisions in Medicine, Michigan State University Press, East Lansing, MI, 1974.

Callahan, Daniel, Abortion: Law, Choice and Morality, The Macmillan Co., New York, NY, 1970.

Campbell, Alastair V., Moral Dilemmas in Medicine, The Williams and Wilkins Co., Baltimore, MD, 1972.

Charles, Cheryl and Samples, Bob (eds.), National Council for the Social Studies, Bulletin 57, Washington, DC 1978.

Ehrlich, Paul R., and Anne H. Ehrlich, Population Resources Environment, W. H. Freeman and Co., San Francisco, CA, 1972.

Fletcher, Joseph, Morals and Medicine, The Beacon Press, Boston, MA, 1960.

Fowler, K. M., Hunger: The World Food Crisis, NSTA Publication, Washington, DC, 1978.

Frankena, William A., Ethics (2nd ed.), Prentice-Hall, Inc., Englewood Cliffs, NJ, 1973.

Goodfield, June, Playing God, Random House, New York, NY, 1977.

Granfield, David, The Abortion Decision, Doubleday and Co., Inc., Garden City, NY, 1969.

Gutcheon, Beth R., Abortion: A Woman's Guide, Abelard-Schuman, Ltd., New York, NY, 1973.

Handler, Philip (ed.), Biology and the Future of Man, Oxford Book Co., New York, NY, 1970.

Hardin, Garrett, Population, Evolution, and Birth Control (2nd ed.), W. H. Freeman and Co., San Francisco, CA, 1969.

Hunt, Robert, and John Arras, Ethical Issues in Modern Medicine, Mayfield Pub. Co., Palo Alto, CA, 1977.

Karp, Laurence E., Genetic Engineering: Threat or Promise?, Nelson-Hall, Chicago, IL, 1977.

Kieffer, George H., Bioethics: A Textbook of Issues, Addison-Wesley Pub. Co., Reading, MA, 1979.

Kohl, Marvin (ed.), Beneficient Euthanasia, Prometheus Books, Buffalo, NY, 1975.

Kubler-Ross, Elizabeth, On Death and Dying, The MacMillan Co., New York, NY, 1969.

Kubler-Ross, Elizabeth, Death: The Final Stage of Growth, Prentice-Hall, Inc., Englewood Cliffs, NJ, 1969.

Lyons, Catherine, Organ Transplants: The Moral Issues, The Westminster Press, Philadelphia, PA, 1970.

Maguire, Daniel C., Death by Choice, Doubleday and Co., Inc., Garden City, NY, 1974.

Mann, Kenneth, Deadline for Survival: A Survey of Moral Issues in Science and Medicine, Seabury Press, New York, NY, 1970.

Manning, Diana H., Society and Food: The Third World, Butterworth, Boston 1977.

Mendelsohn, Everett, et al., Human Aspects of Biomedical Innovation, Harvard University Press, Cambridge, MA, 1971.

National Academy of Sciences, Research with Recombinant DNA, Academy Forum, Washington, DC, March 7-9, 1977.

Neal, Robert E., The Art of Dying, Harper and Row Publishers, Inc., New York, NY, 1973.

Nelson, James B., Human Medicine, Augsburg Publishing House, Minneapolis, MN, 1973.

Nolan, William A., Spare Parts for the Human Body, Random House, Inc., New York, NY, 1971.

Noona, John T., Jr. (ed.), The Morality of Abortion: Legal and Historical Perspectives, Harvard University Press, Cambridge, MA, 1970.

Options, A Study Guide to Population and the American Future, The Population Reference Bureau, Inc., 1755 Massachusetts Ave. N.W., Washington, DC 20036.

Planned Parenthood of New York City, Abortion: A Woman's Guide, Abelard-Schuman, Ltd., New York, NY, 1973.

Potter, Van Rensselaer, Bioethics: Bridge to the Future, Prentice-Hall, Inc., Englewood Cliffs, NJ, 1971.

Ramsey, Paul, Fabricated Man: The Ethics of Genetic Control, Yale Univ. Press, New Haven, CT, 1970.

Ramsey, Paul, The Patient as Person, Yale University Press, New Haven, CT, 1970.

Ramsey, Paul, The Ethics of Fetal Research, Yale University Press, New Haven, CT, 1975.

Riker, A. P., and Riker, C., Finding My Way, Chas. A. Bennett Co., Inc., Peoria, IL 1979.

Rogers, Michael, Biohazard, Alfred A. Knopf, Inc., New York, NY, 1977.

Russell, O. Ruth, Freedom to Die: Moral and Legal of Euthanasia, Human Sciences Press, New York, NY, 1975.

Shore, Milton F., and Stuart E. Golann (eds.), Current Ethical Issues in Mental Health, National Institute of Mental Health, Rockville, MD, 1973.

Stine, Gerald James, Biosocial Genetics: Human Heredity and Social Issues, The Macmillan Co., Inc., 1977.

The Teaching of Bioethics, Report of the Commission on the Teaching of Bioethics, The Hastings Center, Hastings-on-Hudson, NY, 1976.

Vaux, Kenneth, Biomedical Ethics, Harper and Row Publishers, Inc., New York, NY, 1974.

Vaux, Kenneth, Who Shall Live?, Fortress Press, Philadelphia, PA, 1970.

Veatch, Robert M., Death, Dying and the Biological Revolution: Our Last Quest for Responsibility, Yale University Press, New Haven, CT, 1976.

Visscher, Maurice B. (ed.), Humanistic Perspectives in Medical Ethics, Prometheus Books, New York, NY, 1972.

Wade, Nicholas, The Ultimate Experiment, Walker, New York, NY, 1977.

Walbert, David F., and J. Douglas Butler (eds.), Abortion, Society and the Law, The Press of Case Western Reserve University, Cleveland, OH, 1973.

Wertz, Richard W., Readings on Ethical and Social Issues in Biomedicine, Prentice-Hall, Inc., Englewood Cliffs, NJ, 1973.

Wilson, Jerry B., Death by Decision: The Medical, Moral, and Legal Dilemmas of Euthanasia, The Westminster Press, Philadelphia, PA, 1975.

Wogaman, J. Phillip (ed.), The Population Crisis and Moral Responsibility, Public Affairs Press, Washington, DC, 1973.

Wojcik, Jan, Muted Consent: A Casebook in Modern Medical Ethics, Purdue University Press, West Lafayette, IN, 1978.

Young, David P., A New World in the Morning: The Biopsychological Revolution, The Westminster Press, Philadelphia, PA, 1972.

Zumbro Valley Medical Society, Religious Aspects of Medical Care, Zumbro Medical Society, Rochester, MN, 1975.

Some Additional Values Education Materials:

Barman, Charles R., "Integrating Value Clarification With High School Biology," American Biology Teacher, 37(3):150-153, March 1975.

Barman, Charles R., et al., "Moral Development and Its Implications for Biology Teaching," American Biology Teacher, 38(7):436-439, Oct. 1976.

Barman, Charles R., "Four Values Education Approaches for Science Teaching," American Biology Teacher, 42(3): 152-156, March 1980.

Barman, Charles R. and John J. Rusch, "A Case for Bioethics in the Secondary Curriculum," School, Science and Mathematics, 78(1): 3-8, January, 1978.

Hall, Brian P., Value Clarification as Learning Process: A Sourcebook, Paulist Press, New York, NY, 1973.

Hall, Brian P., Value Clarification as Learning Process: A Guidebook, Paulist Press, New York, NY, 1973.

Harmin, Merrill, et al., Clarifying Values Through Subject Matter, Winston Press, Inc., Minneapolis, MN, 1973.

Herron, J. Dudley, "Implicit Curriculum--Where Values Are Really Taught," The Science Teacher, 44(3):30-3l, March 1977.

Hurd, Paul D., "Science, Technology, and Society: New Goals for Interdisciplinary Science Teaching," The Science Teacher, Vol. 42, No. 2, February 1975, pp. 27-30.

Mattox, Beverly A., Getting It Together, Pennant Press, San Diego, CA, 1975.

Morrison, Eleanor S., and Mila Underhill Price, Values in Sexuality, Hart Publishing Co. Inc., NY, 1974.

Piburn, Michael D., "Values Education: Insights from the Social Sciences," The Science Teacher, 43(3):18-20, March 1976.

Raths, Louis E., et al., Values and Teaching, Charles E. Merrill Publishing Co., Columbus, OH, 1966.

Simon, Sidney B., et al., Values Clarification, Hart Publishing Co., Inc., New York, NY, 1972.

Stapp, William B., and Dorothy A. Cox, Environmental Education Activities Manuals (1 through 6), Thomson-Shore, Inc., Dexter, MI, 1974.

Superka, Douglas P., et al., Values Education: Approaches and Materials, Social Science Education Consortium, Inc., Boulder, CO, 1975.

INDEX

TEACHER NOTES

TEACHER NOTES

DATE DUE
